GREENER PRODUCTS

PRODUCTS

The Making and Marketing of Sustainable Brands

SECOND EDITION

GREENER PRODUCTS

The Making and Marketing of Sustainable Brands

SECOND EDITION

Al Iannuzzi

CRC Press
Taylor & Francis Group
Boca Raton London New York

CRC Press is an imprint of the
Taylor & Francis Group, an **informa** business

CRC Press
Taylor & Francis Group
6000 Broken Sound Parkway NW, Suite 300
Boca Raton, FL 33487-2742

© 2018 by Taylor & Francis Group, LLC
CRC Press is an imprint of Taylor & Francis Group, an Informa business

No claim to original U.S. Government works

Printed on acid-free paper

International Standard Book Number-13: 978-1-138-62629-4 (Hardback)
978-1-138-74689-3 (Paperback)

Library of Congress Cataloging-in-Publication Data

Names: Iannuzzi, Al, author.
Title: Greener products : the making and marketing of sustainable brands / Al Iannuzzi.
Description: Second edition. | Boca Raton, FL : CRC Press, 2017. | Includes bibliographical references and index.
Identifiers: LCCN 2017010260 | ISBN 9781138626294 (hardback : alk. paper) | ISBN 9781138746893 (paperback : alk. paper)
Subjects: LCSH: Green marketing. | Green products.
Classification: LCC HF5413 .I16 2017 | DDC 658.8/02--dc23
LC record available at https://lccn.loc.gov/2017010260

Visit the Taylor & Francis Web site at
http://www.taylorandfrancis.com

and the CRC Press Web site at
http://www.crcpress.com

Contents

Section I   The Case for Greener Products

Section II Making Greener Products

Section III   Marketing Greener Products

Foreword

Continuing population growth, a growing middle class around the world, with expanded economic power—all chasing the dream of unobstructed consumerism—is placing impossible demands on our planet's natural capital. We use more energy, consume more resources, and create more waste than ever before. In fact, we are currently "spending at a deficit"—using more natural resources to meet the demands of consumption worldwide than our planet can replace each year.[*] We have no choice but to address the impact that our current models of consumption have on the natural world. And as a key driver of the system, brands must lead the way by making their products more sustainable.

The good news is that in this mandate lies opportunity as well. The products we buy impact not only our own health but also the health of our families, our communities, and the world. Broadly, consumers have higher expectations than ever of the companies they buy from, and this is impacting the brands they choose. Those brands which honestly seek to understand and improve on the direct and indirect impact they have on all their stakeholders—not just their customers and shareholders—will be the ones that take the lead and stand the test of time in the 21st century.

Brands don't have to be perfect, but they *do* need to be authentic and transparent about their impacts, and they *do* need to show continuing improvements in the way they are addressing their environmental and social impact. We have a long way to go to get to a fully sustainable economy—and it will take a commitment from all parts of the economic system, including consumers, to get there. However, the degree to which a company aligns a meaningful brand promise with the way it designs, builds, and delivers its brand to market is key. And business leaders *do* have an opportunity to drive sales and build customer loyalty and brand value by positively addressing the social and environmental challenges we face through purpose-driven product innovation.

Building upon the first edition of *Greener Products*, Al Iannuzzi and his contributors make the case for why it is an imperative to develop and market more socially and environmentally smart brands. Through reviewing some of the many case studies of leading companies' greener product development and marketing programs, readers can glean best practices of smart business strategy and sustainable innovation. *Greener Products* makes it clear: Sustainability doesn't have to come at the sacrifice of profit.

---

[*] Footprint Network. http://www.footprintnetwork.org/our-work/earth-overshoot-day/ (Accessed June 2, 2017).

If you want to know how to make your brand more sustainable and how to sell your better brand to your customers, *Greener Products* explains the details. As you'll experience from reading this book, companies that use sustainability as a driver of innovation and then smartly market their benefits will win. As Al says, when you "hit the *sweet spot,* of having a truly greener product that is communicated in an appropriate way, *everyone wins.*"

KoAnn Vikoren Skrzyniarz
Founder/CEO
Sustainable Brands Worldwide

Preface

The world's resources are under more pressure than ever before. We are using more energy and creating more waste and more air pollution than ever before. There is a growing world population and more importantly a growing middle class in developing nations such as Brazil, Russia, India, and China. With the increased wealth, there is a desire to have products that more developed nations have: cars, computers, mobile phones, televisions, iPads, etc. More of these products means more pollution, increased greenhouse gases, more resources extracted, and more waste to be disposed of. But the even bigger question is: Where will all of these raw materials to come from? Regardless of how dire you feel things are regarding environmental issues and the future of the world, we all can agree that there is going to be more competition over raw materials to make the products that the growing middle-class desires.

At the very least, competition for resources will make products much more expensive to make and, in a worst-case scenario, would severely restrict our ability to provide these products due to a dearth of raw materials. Considering this forecast, business as usual is not an option. Institutionalizing sustainability in new product-development processes will be an imperative. Businesses must respond and do things differently, evaluate life cycles and minimize use of raw materials, and consider impacts during customer use and product end of life. Deploying such concepts as design for the environment, green chemistry, and biomimicry needs to be commonplace and built into the way new products are brought to market. This is the reason I authored this book, to offer guidance on how to make products greener or more sustainable and how to appropriately market them.

Before the first edition of *Greener Products*, I was unaware of any books that combined the concepts of design for the environment/industrial ecology and green marketing into one book, and I still believe that this book is the only one that combines these concepts. Whenever I speak about greener products, there are two things I usually say:

1. *There is no such thing as a green product.*
2. *What good is a greener product if no one knows about it?*

The reason for these assertions is that life-cycle assessments have shown that every product has impacts, from raw materials to transportation, manufacturing, customer use, and end of life. Every product can be improved in some way, which is why I use the term Greener. And once you have a greener product, you must appropriately inform your customers why it is greener;

if you don't do this correctly, it is essentially worthless to have an improved product since no one will know about it. That's why it is critical to *make and market greener products*—the communication aspect is just as important as *having* a greener product.

The words we use are important, and in this book, I use the terms greener products and more sustainable products synonymously. These terms are descriptors for products that demonstrate improvements in environmental or social areas. I should also mention that I don't believe that any corporation is *sustainable* or is *a sustainable business*. In my mind, sustainability is a journey and to the point above, every product and every corporation have an impact. There are a lot of things we can do to minimize our impact, but we can't eliminate them completely, that is why I see sustainability as a journey. I believe that what customers all over the world expect from responsible companies is to know that they are moving in the right direction, keeping an eye on sustainability, and making their product's footprint smaller and smaller.

This book has three main sections:

Section I: The Case for Greener Products
Section II: Making Greener Products
Section III: Marketing Greener Products

Why is it an imperative to bring greener products to market? This question is answered in Chapters 1 through 3. We get a deeper understanding of the ecosystem pressures on the Earth and explore the market pressures being placed on businesses. This includes customers who buy products in supermarkets, business customers, and government purchasers who created a big demand of manufactures to bring more sustainable brands to market. People throughout the world are more health conscious and are hearing new stories about the effects of chemicals on their health; therefore, they are seeking more organic and sustainably sourced products.

Global companies such as Walmart, Tesco, Lowe's, and IKEA, among others, have embraced sustainability and are triggering other companies to build eco-innovation into their product offerings as well. Governments have responded by putting regulatory pressure on product developers to initiate tougher new requirements such as removing materials of concern (e.g., PVC, triclosan, brominated flame retardants, BPA) and design more recyclable packaging, and to facilitate product take-back at its end of life. These global regulations started out in Europe and have exploded throughout the world, making it very complicated for businesses to keep up with the ever-changing requirements and be compliant.

How do you make greener products? To find out how to go about making a greener product, it's a good idea to evaluate the best practices from leading companies. Chapter 4 covers best practices in a variety of industrial sectors, such as apparel, chemicals, electronics, consumer packaged goods, medical products, and energy.

I am privileged to have one of the leading thinkers and practitioners in developing greener products, Jim Fava, share his thoughts on frameworks used to make products more sustainable in Chapter 5. Jim and his associates share leading practices and discuss lessons learned from consulting companies on how to address the myriad demands on companies noted above.

A new addition to *Greener Products* is Chapter 6 by Libby Bernick, a thought leader in the burgeoning science of Natural Capital. Libby sets out to explain why we need to think about the costs to nature of the products we manufacture. There are some examples of companies that have used natural capital to make decisions on improving their product's sustainability performance and they are illustrated in the chapter.

A greener product is useless if no one knows it exists. Appropriate marketing of eco-improved products is a critical aspect of a sustainability program. Why should we consider green marketing? It's the secret sauce that brings greener products to life. The third section of this book addresses *green marketing*—our evaluation of sustainable brand marketing begins by reviewing some exemplary case studies in Chapter 7. We look at some of the best examples of green marketing from companies that have been very successful in making inroads with consumers.

Consumer behavior data indicate that all over the world, there is a desire to buy products that are made by responsible companies and have greener attributes. But, they don't want to pay more for them. So if companies want to be successful in selling greener products, they cannot be at a higher price point, and they should sell them with an "and" in mind—an effective product at the right price *and* it's greener.

Not only must consumer marketers be concerned about bringing eco-innovative products to market but also business-to-business (B2B) marketers must be savvier in greener product offerings. Scorecards by Walmart, Kaiser Permanente, and Procter & Gamble and the advent of B2B green purchasing make the case that all product marketers must pay attention to making and marketing greener products.

Based on my analysis of successful green marketing campaigns, there are three keys to winning:

1. Have a credible greener product story
2. Meet your customers' greener product demands
3. Appropriately communicate the products' greener attributes

These three characteristics will be used to evaluate the approach of several leading companies in business-to-consumer (B2C) and B2B marketing. Some of their initiatives have been ground-breaking and have paved the way to the current state of green marketing. We evaluate what has made Ecomagination so impactful, how Green Works changed the game for mainstream green marketing, as well as innovative approaches by Honest Tea and Neutrogena Naturals.

Understanding a good framework for sustainable brand marketing from a leading marketing and communications firm, OgilvyEarth, will be helpful to anyone interested in marketing green. That is exactly what you will find in Chapter 8. There are specific steps that can be taken to enable the marketer to have a successful campaign and examples and case studies of effective company programs.

There are certain elements that can make a marketing campaign more effective and certain things that will blow it up. That's what Chapter 9 explores; helpful elements like the use of eco-labels are reviewed, along with understanding regulatory requirements. There are key aspects that must be understood such as the FTC green guides and the UK DEFRA green marketing rules. Perhaps, even more important than the regulatory requirements is the avoidance of greenwashing. Inappropriate use of marketing claims can be a death blow to a brand—it's critical to avoid greenwashing; evaluating The Seven Sins of Greenwashing will help. Several methods for marketing the right way, to enhance and protect your brand, are discussed.

Finally, Chapter 10 concludes with an evaluation of the best practices that leading companies have for making and marketing greener products. Through an evaluation of the various company initiatives discussed in the Making Greener Products section of the book, we see that there are some commonalities among the companies that are best at developing eco-improved products. Use of all or a combination of these activities will enable any company to bring more sustainable products to the market.

A review of the common approaches of successful green marketing campaigns makes it clear to marketers what elements should be considered to have a successful approach to reaching customers. After all, what's the good of making improvements to a product if you don't effectively communicate with your customers?

Making and marketing greener products is no longer a nice *to do*—it is a business imperative. I believe this book makes it clear that the world needs greener products and it's up to the marketers, R&D, and product stewardship leaders to make that happen. No matter what industry you're in, it makes business sense to institutionalize these concepts. Evaluating the leading practice examples and techniques in this book will give any product developer or marketer some ideas that they then can translate into their own company culture.

I am hopeful that this book will not only benefit business but will also enable students to better understand what it takes to make and market a greener product and will also be useful for academics and NGOs. We all are concerned about becoming more sustainable, and I believe that it is possible to strike a balance in meeting the world's product needs while appropriately reducing unsustainable practices to meet this demand.

Al Iannuzzi

Disclaimer

The views and opinions expressed in this book are those of the author and do not necessarily reflect the official policy or position of Johnson & Johnson.

Acknowledgments

I am truly privileged to have the opportunity to write this book. My purpose for doing it was to encourage companies to make the world a better place and to demonstrate how others have done it through making and marketing truly greener products. Writing a book takes a lot of effort, and it's particularly challenging when you're working full time, but I am extremely thankful for having an exceptional team that helped and contributed to what I hope will be a book that makes a difference.

First, I must thank my family whom I love immensely. Before I start anything that will be time-consuming, I request permission from the most important person in my life, my wife Ronnie, who doubled as a proofreader and advisor on this project, so thank you, Ronnie, for the approval, understanding, and help. Next, I would like to thank my daughter Marissa, who helped with some of the research and writing for this book—I appreciate you bunches. To my daughter Gianna and son Alphonse, thank you for your understanding and patience.

Having a super intern from Indiana University, where I am an adjunct professor, was a huge blessing, so I thank you Virginia Miller for the great work on Chapter 1, and all of the other work you did—you have a bright future ahead of you.

Another big boon was the fantastic thought leaders who contributed to this effort. Jim Fava and the crew from Anthesis Group shared their experiences from many years of sustainability consulting: Lauren Bromfield, John Heckman, Doug Lockwood, Chris Peterson, Andrea Smerek, and Allen Spray. Jim Fava has been a long-time trusted advisor and a catalyst in the world, encouraging companies to be more sustainable.

Libby Bernick did a wonderful job with the chapter on Natural Capital and demonstrated her leadership in this quickly developing area. Having some expert communicators/marketers from Ogilvy was an added benefit to Greener Products. John Jowers and Ivellisse Morales provided excellent clarity on how to effectively market your product using proven techniques from their experience at OgilvyEarth.

Finally, I would like to thank Taylor & Francis for supporting me in a second edition of *Greener Products*, which will encourage others to pursue making their brand more sustainable.

We all are stewards of the earth—let's make the world better with Greener Products.

> The heavens are Yours, the earth also is Yours; The world and all it contains, You have founded them. (Psalm 89:11)

> Moreover, it is required in stewards that one be found faithful. (1 Corinthians 4:2)

Author

Al Iannuzzi, PhD, is senior director in the Worldwide Environment, Health, Safety & Sustainability (EHS&S) Department at Johnson & Johnson. He leads the Global Product Stewardship program, and is a member of the Worldwide EHS&S Leadership Team. He has over 30 years' experience in the EHS field and has developed and directs Johnson & Johnson's Earthwards® greener product design process. Prior to J&J, Al worked for the NJ Department of Environmental Protection and at an environmental consulting firm. Al received his PhD in Environmental Policy from the Union Institute & University in Cincinnati, Ohio, where he researched EHS self-regulation programs. He is the author of the books, *Greener Products: The Making & Marketing of Sustainable Brands* (CRC Press, 2011) and *Industry Self-Regulation and Voluntary Environmental Compliance* (CRC Press, 2002) and has written numerous articles and blogs on product stewardship and environmental compliance. Al is an adjunct professor at Indiana University where he teaches on product improvement and sustainability. He has maintained the certification as a qualified environmental professional since 1996 and speaks regularly at sustainability conferences and universities.

Section I

The Case for Greener Products

1

Introduction

Virginia Miller

Things Will Never Be the Same

It has been a long steady climb, but it seems sustainability has finally reached a tipping point. We are entering a new era of resource management and **things will never be the same**. The rise of the global middle class is occurring at a rapid clip, with an additional 3 billion people expected to enter the consumer market by 2030 (Nyquist et al. 2016). The growing urban middle class brings with it an unprecedented demand for resources, putting more strain on the global supply of raw materials, fossil fuels, food, and water. Furthermore, greenhouse gas emissions and waste generation will continue to soar as global consumption increases. With this new reality apparent, **"business as usual" is no longer an option**. Resources will have to be used in a more efficient way in order to meet the needs of global population, both present and future.

Under the pressure of these growing demands, the quest for sustainability has transformed the business landscape. Sustainability that was once viewed as a simple risk-management function is now regarded as a favorable business opportunity. The growing demand for resources offers a chance for business leaders to meet demand in a sustainable way, and in turn revolutionizes how we think about consumption. Businesses are even shifting their missions and values to align with their sustainability goals, with over a third putting it in their top three objectives (McKinsey and Company 2014). Greener products are playing a bigger role in business than ever before. Sustainability is no longer some idealistic notion; it is at the forefront of development. Company leaders are rallying behind sustainability, and **things will never be the same**.

What Caused This Shift?

Environmentalism has appeared sporadically throughout history, in many different forms, all over the world. Many associate the beginning of the modern environmental movement with Earth Day and the legislative fervor of the 1970s, although its origins are rooted a bit more deeply in the past. "In wildness is the preservation of the world," wrote Henry David Thoreau in his nature-themed essay *Walking*. Thoreau and the transcendentalists of the mid-1800s were some of the first to suggest preserving the natural world for its beauty and potential for spiritual enlightenment, not to mention its immense practical value. These philosophers realized that humans rely heavily on nature for their survival, and the nation's natural bounty was not infinite. The 1860s brought the naturalist John Muir and the creation of the first national park. The National Park Service was created soon after, in 1916—recently celebrating its 100th anniversary. Although "going green" may seem like a recent trend, environmental consciousness is deeply woven into the fabric of American history, and the concept is, arguably, quintessentially American.

In the wake of World War II, environmentalism quickly took a backseat to growth and innovation. The postwar economic boom brought with it increased consumption, waste, and pollution. The pollution continued largely unchecked until the 1960s brought a new wave of environmentally conscious thinking, catapulted by Rachel Carson's landmark book *Silent Spring*, which helped set the stage for the environmental movement by exposing the environmental hazards of pesticides and by questioning humanity's unwavering faith in technological advancement. In June 1972, the United Nations Conference on the Human Environment was held in Stockholm, Sweden. This conference would lay the foundation for modern global environmental policy. The 1970s also brought a rapid series of new environmental legislation, the creation of the U.S. Environmental Protection Agency, and the inaugural celebration of Earth Day. Thus began the consumer-driven environmental movement of today. It rapidly gained momentum over the following decades and exploded onto the world business scene.

The recent shift toward more sustainable products is due, in part, to greater awareness of major environmental issues such as climate change and consumers wanting to make a difference with their purchasing decisions (Figure 1.1). 2016 is now the hottest year on record (NASA 2017). Awareness of this crucial issue is certainly high; yet, it is far from the only concern on the minds of executives and corporate leaders.

According to the United Nations, a staggering 92% of the world's population lives with air pollution more than the recommended limits. Although tremendous gains have already been made, 663 million people are still without improved drinking water, and 2.4 billion lack access to improved sanitation. Water scarcity already affects 40% of the world's population and is projected to rise. "Sustainability is a worldwide concern that continues to

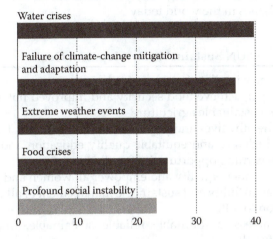

The top five global risks of highest concern for the next 10 years percent of respondents

FIGURE 1.1
The top five global risks of highest concern for the next 10 years. (From Boumphrey, Sarah, 2016, Sustainability and the New Normal for Natural Resources. *Euromonitor International*. p. 15.)

gain momentum—especially in countries where growing populations are putting additional stress on the environment," says Grace Farraj, senior vice president, Public Development and Sustainability, Nielsen. Consumers in these developing markets—such as Latin America, Asia, Middle East, and Africa—are often closer to and more aware of the problems in their surrounding communities, and thus are very likely to seek out and pay a premium for sustainable offerings (Nielsen 2015a).

To tackle these global issues, world leaders assembled at the United Nations Sustainable Development Summit in September 2015 to adopt the 2030 Agenda for Sustainable Development, which includes a set of 17 Sustainable Development Goals (SDGs; Figure 1.2), effectively replacing the

FIGURE 1.2
Sustainable Development Goals. (From United Nations 2016c.)

highly successful Millennium Development Goals that were in place from 2000 to 2015. These goals are targeting the most pressing environmental and humanitarian issues in the world today.

UN Sustainable Development Goals

1. End poverty in all its forms everywhere
2. End hunger, achieve food security and improved nutrition, and promote sustainable agriculture
3. Ensure healthy lives and promote well-being for all at all ages
4. Ensure inclusive and equitable quality education and promote lifelong learning opportunities for all
5. Achieve gender equality and empower all women and girls
6. Ensure availability and sustainable management of all water and sanitation for all
7. Ensure access to affordable, reliable, sustainable, and modern energy for all
8. Promote sustained, inclusive, and sustainable economic growth; full and productive employment; and decent work for all
9. Build resilient infrastructure, promote inclusive and sustainable industrialization, and foster innovation
10. Reduce inequality within and among countries
11. Make cities and human settlements inclusive, safe, resilient, and sustainable
12. Ensure sustainable consumption and production patterns
13. Take urgent action to combat climate change and its impacts
14. Conserve and sustainably use the oceans, seas, and marine resources for sustainable development
15. Protect, restore, and promote sustainable use of terrestrial ecosystems, sustainably manage forests, combat desertification, halt and reverse land degradation, and halt biodiversity loss
16. Promote peaceful and inclusive societies for sustainable development, provide access to justice for all and build effective, accountable, and inclusive institutions at all levels
17. Strengthen the means of implementation and revitalize the global partnership for sustainable development

---

The SDGs have stirred up significant activity; government leaders worldwide have seemingly embraced the new goals, with numerous presidential decrees, national action plans, new policies, budgets, and stakeholder collaboration platforms being rolled out in direct response to the SDGs. Businesses have also taken notice, and many are incorporating the SDGs into their objectives with surprising seriousness (Atkisson 2016). The Business and Sustainable Development Commission, co-founded by Unilever CEO Paul Polman in 2016, is bringing business leaders together in support of the SDGs. The commission's aim is "to inspire business leaders to seize upon sustainable development as the greatest economic opportunity of a lifetime, and to accelerate the world's shift to inclusive growth" (Business & Sustainable Development Commission 2016). The commission will present a report in a year's time outlining new business and financial models, as well as market opportunities for companies that are investing in sustainable approaches (Unilever 2016).

Our aim is to inspire business leaders to seize upon sustainable development as the greatest economic opportunity of a lifetime, and to accelerate the world's shift to inclusive growth. (Business and Sustainable Development Commission)

Constant Pressures on the Environment

A constant barrage of environmental catastrophes is being witnessed by the world, and the stories never stop coming. Just recently, in October 2015, researchers detected the largest natural gas leak in the US history in southern California. Almost 100,000 tons of methane poured into the atmosphere before it was permanently sealed on February 18, 2016, almost 4 months after it was detected. The impact of this gas leak on climate change is said to be the equivalent of the annual emissions of half a million cars, having a far bigger warming effect than the BP oil spill in the Gulf of Mexico in 2010 (McGrath 2016).

This was not the only environmental disaster that year. On November 5, 2016, a mining dam ruptured near the rural village of Bento Rodrigues, Brazil, releasing 62 million cubic meters of toxic mining waste. A wave of toxic waste several meters high surged through the village, killing 18 people, destroying homes and the natural environment. The sludge eventually found its way into the Doce River system and, finally, the Atlantic Ocean. Experts say this is the largest disaster of its kind and will have severe consequences on the surrounding environment for decades (Damsgarrd 2016). Another mine spill occurred in August 2015 near Silverton, Colorado. The Gold King Mine released about 3 million gallons of wastewater laced with heavy metals into Cement Creek, a tributary of the Animas River, turning the entire river bright orange for several days (US EPA 2016). In China, Beijing recently issued its first-ever air pollution "red-alert," the highest warning for air pollution severity, forcing schools and factories to close, and people to remain indoors for days on end (BBC 2015).

There is little doubt that these disasters shape our thinking about the environment and the pressure we are putting on it. Consumers have a tendency to want to punish the corporate violators by not purchasing their products. Global environmental damage caused by human activity was estimated at $6.6 trillion and about 11% of global gross domestic product in 2008. The **top 3,000 public companies are responsible for one third of global environmental damage**; it is no wonder citizens everywhere want to move their buying power toward environmentally enlightened companies and away from those perceived as irresponsible (Environmental Leader 2010). Consumers today are well informed, and 84% of global consumers say they seek out responsible products whenever possible (Sustainable Brands 2016). The desire to "make a difference" with purchases is growing. It is imperative that companies are aware of current environmental and resource issues, and hold themselves accountable for the impact of their products. Companies also need to begin looking beyond the surface, as the impact of products stretches deep into the supply chain. This includes raw materials that are harvested in a plantation that has destroyed the rich biodiversity of a tropical rain forest, to the minerals in your electronics that are mined from a country that is exploiting its citizens.

Environmental Concerns of Shoppers

- The amount of **waste** that's produced
- The damage to **wildlife** and loss of biodiversity
- The amount of **resources** left for future generations
- The impact on the **landscape**
- The impact on **climate change**

Source: Institute of Grocery Distribution, *The environmental concerns of shoppers, 2016*, http://www.igd.com/The_environmental_concerns_of_shoppers_article.

Mainstreaming of Greener Products

Pressure on businesses to "go green" is growing rapidly, but it is not coming from the government. Perhaps surprisingly, the driving force behind more sustainable product design is, in fact, market pressure! A prime example is the world's largest retailer Walmart. Once Walmart started asking for more sustainable merchandise, it began a boon for greener products. Being such a large retailer, its demands for greener products were momentous and far-reaching and the effects stretched deep into the supply chain as well as catalyzed the competition.

Once Walmart started their sustainability journey, many other companies followed suit by setting their own sustainability goals. Competition ensued between rival companies, further pushing the boundaries of sustainable design. Due to the emerging demand, sustainability had become a key driver of innovation. More sustainable products have cropped up in all aspects of business, trying to address the consumer's desire for more ethical goods. In turn, they provide an additional benefit, such as providing safer products for the home or saving money through energy efficiency.

Once Walmart started asking for more sustainable merchandise, it began a boon for greener products. Being such a large retailer, its demands for greener products were momentous and far-reaching.

Sustainability market pressures have been affecting all industrial sectors. An obvious illustration is how the automotive industry has been revolutionized with the success of the **Toyota Prius** hybrid. Being the first car that has had a green platform from its inception, it has been well received by a public that is eager to purchase a high fuel-efficiency product. Though its success was likely due in part to the rapidly rising gas prices of the early 2000s, the sale of electric and hybrid vehicles is still on the rise in both developing and developed countries, despite a slump in oil prices (Kaye 2016). Since the Prius, many major car manufacturers have released their own version of hybrid vehicles, and more are expected to appear in the up coming years.

Further highlighting this revolution, the highly anticipated **Tesla Model 3** has garnered a lot of attention since it was revealed in April 2016. The so-called "electric vehicle for the masses" hits the sweet spot on price, design, and performance—and people are excited. Over 180,000 preorders were placed in the first 24 hours, with each reservation requiring a hefty $1000 deposit. Where traditional automakers typically design eccentric, unconventional electric vehicles marketed toward environmentalists (think Nissan Leaf), Tesla's "think-big" attitude may have resulted in the first true "everyman's" electric car—and the outlook is good. A recent automotive forecast by McKinsey & Company estimates that by 2030, electric vehicles could represent about 30% of all new cars sold globally, and up to 50% of those sold in China, the European Union, and the United States (Nyquist et al. 2016). All this is happening while oil prices are at their lowest levels in over a decade!

Beverage behemoths **Coca-Cola** and **PepsiCo** have also jumped on the sustainability bandwagon. Water is the primary ingredient in their beverages, and both companies recognize that the increasing stress on water supply is a risk to their business. In August 2016, Coca-Cola announced that it had replenished just under 200 billion liters of water to the environment and communities, over 115% of the water used in its beverages last year. This achievement makes it the first Fortune 500 company to replenish all of the water it consumed. They accomplished this through 248 community water partnership projects in 71 countries. Some projects are focused on returning water directly to the source, while others are contributing to communities where there is a pressing need. Meanwhile, PepsiCo has reduced its operational water use by 26% compared to 2006, exceeding its goal to reduce its water use by 20%. These water conservation efforts over the past 5 years have saved the company $80 million. At the same time, PepsiCo Foundation partnerships have helped provide safe drinking water to 9 million people since 2006 (Sustainable Brands 2016). Additionally, Gatorade, a PepsiCo product, has released a certified organic version of the world's best-selling sports drink. G Organic contains only seven ingredients, none of which are artificial. Every step of the process was approved by the USDA, which aims to ensure organic products are more natural and less harmful to the environment.

The largest furniture retailer in the world **IKEA** has launched an immensely successful sustainability campaign, with a 2020 goal of producing more energy than it uses. The results so far are promising. IKEA owns 157 wind turbines with a capacity of 345 megawatts and has installed over 550,000 solar panels, totaling 90 megawatts. Their extensive renewable energy portfolio rivals that of some energy companies (Kelly-Detwiler 2014). The furniture giant sells everything from appliances to textiles to bicycles. They have already switched 100% of their lighting offerings to LEDs. The entire range of appliances they offer are energy efficient, including induction technology on stovetops, A+++ rated refrigerators that use 40% less energy, and water taps that reduce consumption by 30%–50%. Furthermore, they aim to offer these products **at an affordable price**, making sustainable living further available to the general public.

Another major example of the mainstreaming of greener products is the green building movement. There has been a large shift in recent years to energy-efficient housing. In the United States, buildings account for 38% of all CO_2 emissions, 13.6% of all potable water use, and 73% of electricity use (USGBC 2016). Green buildings are continuing to make up a larger share of the housing market, appealing both to those who are concerned about the environmental impacts of energy use, as well as those who just want to save on energy bills. One major reason for the propagation of green buildings is the creation of the Leadership in Energy and Environmental Design (LEED) certification system.

LEED is the most widely used third-party verification system for green buildings, with around 1.85 million square feet of building space being certified on a daily basis. The company's goal is to "change the way we think about how buildings and communities are planned, constructed, maintained and operated." Buildings that are LEED certified have improved environmental performance in areas such as energy use, water efficiency,

CO_2 emissions reduction, improved indoor air quality, and use of recycled content materials, sustainably sourced materials, and stewardship of resources (USGBC 2016).

Many people are gaining independence from the power grid completely with the advent of "net-zero" homes that are capable of generating their own electricity, predominantly through the use of solar panels. These new homes are so energy efficient that their electricity consumption is essentially zero. It is not uncommon for energy consumption to be so low that the electricity meter actually runs backward, selling electricity back to the grid. Materials and construction costs do increase the price of clean-energy homes, but not as much as you might think—and costs are dropping. Green homes typically cost between 5% and 10% more to build than a similar conventional home, seemingly negligible once you consider the energy savings, tax credits, and added resale value (Rohwedder 2013).

Why the Focus on Greener Products?

Annie Leonard makes a very good case for focusing on how we create, use, and dispose of products in her short documentary, *The Story of Stuff.* Despite being almost a decade old at the time of this writing, the video still discusses some of the most relevant issues facing our consumption culture today. After its debut in 2007, it quickly became an Internet sensation with over 40 million views worldwide and spurred an entire series of similar movies and projects promoting social and environmental change. If you have not seen it, I would encourage you to view it at storyofstuff.org. A compelling case is made that products need to be managed in a much more sustainable manner (Story of Stuff 2016).

Traditionally, environmental management practices only focused on managing risks and reducing the carbon footprint at the manufacturing and production level. With the emergence of life-cycle thinking (also known as cradle to grave), we are now looking at the environmental, social, and economic impacts of a product over its entire life cycle. In fact, some of the greatest environmental benefits occur through the selection of raw materials, packaging, distribution, use phase, and disposal of a product.

One of the greatest product examples of life-cycle thinking is cold water laundry detergent, developed by Proctor & Gamble. Life-cycle assessments revealed that the heating of water during its use in the home consumed far more energy than any other life-cycle stage. In response, a detergent that functions in cold water was introduced into the market, which does not require heating of wash water (Janeway 2016).

Circular Economy

One of the newest sustainability concepts is that of a circular economy, initiated by the Ellen MacArthur Foundation (Figure 1.3). The circular economy takes the idea of life-cycle thinking one step further; the current "take, make, dispose" economic model is dependent on an endless supply of cheap raw materials and energy, as well as an unlimited capacity to store waste. This model is rapidly exceeding the physical limits of the planet, and a circular economy offers a viable alternative. A circular economy is "restorative and regenerative" by design, seeking to produce zero waste through an enhanced flow of goods and services (Ellen MacArthur Foundation 2016).

The concept of a circular economy has exploded in popularity, with many companies embracing the idea. Among several other social and environmental initiatives, clothing brand Eileen Fisher, Inc. has incorporated a circular economy into their business model through their line of recycled clothing "Green Eileen." Begun as a nonprofit in 2009, Green Eileen recycles clothing by reselling items in their Green Eileen stores. With the average American throwing out 70 pounds of clothing each year, Green Eileen extends the life of used garments by creating a second market. Anything that cannot be resold is upcycled through a program they call

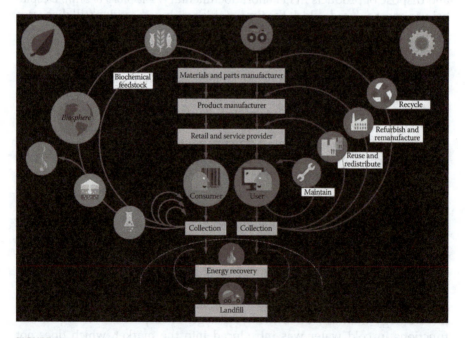

FIGURE 1.3
Ellen MacArthur circular economy. (From Ellen Macarthur Foundation, 2016, Circular Economy, https://www.ellenmacarthurfoundation.org/circular-economy. With permission.)

"Remade in the USA," which turns flawed garments into entirely new designs. The program has been extremely successful, recycling over 600,000 pieces of clothing to date. Many other big-name brands, including Levi Strauss, Nike, Dell, Patagonia, IKEA, and H&M, are working to make the concept of a circular economy a reality, largely through innovations in the recycling of material.

Adidas recently made a splash with their new athletic shoes made almost completely of recycled plastic from oceans. Partnered with Parley for the Oceans, Adidas has released the UltraBOOST Uncaged Parley, with an upper made from 95% ocean plastic collected from the Maldives. The shoelaces, heel cap base, heel webbing, heel lining and the sock liner cover are also made with 100% recycled materials. Although only 7,000 pairs have gone up for sale as of November 2016, Adidas has much bigger plans in the works. The company plans to make at least one million pairs by the end of 2017 and states that their ultimate ambition is to eliminate virgin plastic from their supply chain completely. The short video *From Threat into Thread* tells the story behind the shoes, which can be viewed at http://www.adidas.com/us/parley (Adidas 2016; McAlone 2016).

Things Will Never Be the Same

Sustainability is here to stay, and the movement has already had some monumental victories. In December 2015, 195 countries at the 21st Conference of the Parties of the UNFCC adopted one of the most significant diplomatic agreements in history. The landmark Paris agreement is an international treaty, the first of its kind, to address climate change. On October 4, 2016, the European Union voted to ratify the Paris agreement, passing the 55-country threshold needed for the agreement to enter into force. A huge triumph for the planet, the Paris agreement is a big step in the right direction for worldwide sustainability.

Following the conference in Paris, France became the first country in the world to ban plastic plates, cups, and utensils. The new law, which will go into effect in 2020, is a part of the country's Energy Transition for Green Growth Act. The same legislation banned the use of plastic bags in grocery stores and markets. According to French lawmakers, the goal is to promote a "circular economy" of waste disposal, "from product design to recycling" (McAuley 2016). In December 2015, President Barack Obama signed a bill prohibiting the sale of products containing microbeads. Often used in exfoliating and cleansing products, microbeads have been a huge concern for aquatic environments, where the beads eventually end up (Imam 2015). The UK quickly followed suit, pledging to ban them by 2017 (BBC 2016).

Government regulation is only one piece of the sustainability puzzle. Consumers will continue to drive sustainable innovation, and research shows they are demanding it. The Shelton Group's Eco Pulse study from 2016 states that 45% of Americans say buying/using eco-friendly products is an important part of their personal image. This is a significant finding, as in all previous years that statistic has not surpassed 26% (Enkema 2016). Demand for environmentally friendly products is only expected to grow as the younger generations increase their buying power. Despite the fact that Millennials have grown up in one of the most difficult economic climates in the past 100 years, a 2015 Nielsen global online study found that they are the most willing to pay extra for more sustainable products—an astonishing 75%, up from approximately half in 2014 (Nielsen 2015a). If even in tough economic times we are seeing a growing demand for greener products, what will it be like once the global economy picks up? Although the future of the planet and its resources remain uncertain, one thing is evident. To be successful in the years to come, companies will need to embrace a new culture of sustainability and embed this into new product development processes.

Things will never be the same.

References

Adidas. 2016. Parley. http://www.adidas.com/us/parley (Accessed November 16, 2016).

AtKisson, Alan. 2016. Why the SDGs Are Leading a Global Transformation. Greenbiz. October 3. https://www.greenbiz.com/article/why-sdgs-are-leading-global-transformation (Accessed October 4, 2016).

BBC News. 2015. China Pollution: First Ever Red Alert in Effect in Beijing. December 8. http://www.bbc.com/news/world-asia-china-35026363 (Accessed September 27, 2016).

BBC News. 2016. Plastic Microbeads to Be Banned by 2017, UK Government Pledges. September 3. http://www.bbc.com/news/uk-37263087 (Accessed Sept 5, 2016).

Boumphrey, Sarah. 2016. *Sustainability and the New Normal for Natural Resources. Euromonitor International,* p. 15.

Business & Sustainable Development Commission. 2016. http://businesscommission.org/ (Accessed October 1, 2016).

Damsgarrd, Nina. 2016. The forgotten tragedy: Revisiting Brazil's "Worst Ever" Environmental Disaster. The Argentina Independent. April 7. http://www.argentinaindependent.com/currentaffairs/analysis/revisiting-brazils-worst-ever-environmental-disaster/ (Accessed September 26, 2016).

Ellen MacArthur Foundation. 2016. Circular Economy. https://www.ellenmacarthurfoundation.org/circular-economy (Accessed October 1, 2016).

Enkema, Susannah. 2016. Sustainability and American Identity. Part 1. Shelton Group. August 31. http://sheltongrp.com/sustainability-marketing-and-identity-part-1/ (Accessed September 14, 2016).

Environmental Leader. 2010. Top Public Companies Cause One-Third of the World's Environmental Damage. October 5. http://www.environmentalleader.com/2010/10/05/top-public-companies-cause-one-third-of-the-worlds-environmental-damage/#ixzz4PS2BPHm1 (Accessed October 1, 2016).

Imam, Jareen. 2015. Microbead ban signed by President Obama. CNN. December 31. http://www.cnn.com/2015/12/30/health/obama-bans-microbeads/ (Accessed September 5, 2016).

Institute of Grocery Distribution. 2016. The Environmental Concerns of Shoppers. http://www.igd.com/The_environmental_concerns_of_shoppers_article (Accessed September 14, 2016).

Janeway, Kimberly. 2016. Don't Bother Using Hot Water to Wash Your Laundry. Consumer Reports. August 25. http://www.consumerreports.org/washing-machines/dont-bother-using-hot-water-to-wash-your-laundry/ (Accessed September 1, 2016).

Kaye, Leon. 2016. Despite Low Oil Prices, Electric Vehicle Sales Keep Rising. Triple Pundit. August 24. http://www.triplepundit.com/2016/08/despite-low-oil-prices-electric-vehicle-sales-keep- (Accessed September 1, 2016).

Kelly-Detwiler, Peter. 2014. IKEA's Aggressive Approach To Sustainability Creates Enormous Business Opportunities. Forbes. February 2. sustinhttp://www.forbes.com/sites/peterdetwiler/2014/02/07/ikeas-aggressive-approach-to-sustainability-creates-enormous-business-opportunities/#2db7be804ec8 (Accessed September 13, 2016).

McAlone, Nathan. 2016. Adidas Is Selling Only 7,000 of These Gorgeous Shoes Made From Ocean Waste. Business Insider. November 4. http://www.businessinsider.com/adidas-shoe-from-plastic-ocean-waste-2016-11/#-1 (Accessed November 16, 2016).

McAuley, James. 2016. France Becomes the First Country to Ban Plastic Plates and Cutlery. The Washington Post. September 19. https://www.washingtonpost.com/news/worldviews/wp/2016/09/19/france-bans-plastic-plates-and-cutlery/ (Accessed October 2, 2016).

McKinsey and Company. 2014. Sustainability's Strategic Worth: McKinsey Global Survey Results. July 2014. http://www.mckinsey.com/business-functions/sustainability-and-resource-productivity/our-insights/sustainabilitys-strategic-worth-mckinsey-global-survey-results (Accessed September 8, 2016).

McGrath, Matt. 2016. California Methane Leak "largest in US history". BBC News. February 26. http://www.bbc.com/news/science-environment-35659947 (Accessed September 26, 2016).

NASA. 2017. NASA, NOAA Data Show 2016 Warmest Year on Record Globally. https://www.nasa.gov/press-release/nasa-noaa-data-show-2016-warmest-year-on-record-globally.

Nielsen. 2015a. Consumer-Goods' Brands That Demonstrate Commitment to Sustainability Outperform Those That Don't. October 12. http://www.nielsen.com/ug/en/press-room/2015/consumer-goods-brands-that-demonstrate-commitment-to-sustainability-outperform.html (Accessed November 14, 2016).

Nielsen. 2015b. Green Generation: Millennials Say Sustainability Is a Shopping Priority. November 5. http://www.nielsen.com/us/en/insights/news/2015/green-generation-millennials-say-sustainability-is-a-shopping-priority.html (Accessed August 31, 2016).

Nyquist, Scott, Matt Rogers, and Jonathan Woetzel. 2016. *The Future Is Now: How to Win the Resource Revolution. McKinsey Quarterly*, 4: 100–117.

Rohwedder, Cecilie. 2013. Green Goes Mainstream for New Homes. The Wall Street Journal. May 2. http://www.wsj.com/articles/SB1000142412788732378970457 8443173932450096 (Accessed September 14, 2016).

Story of Stuff. 2016. *From a Movie to a Movement.* http://storyofstuff.org/ (Accessed September 9, 2016).

Sustainable Brands. 2016. Coca-Cola, PepsiCo Exceed Water Stewardship Goals. August 31. http://www.sustainablebrands.com/news_and_views/waste_ not/sustainable_brands/coca-cola_pepsico_exceed_water_stewardship_ goals (Accessed September 6, 2016).

Unilever. 2016. New Global Commission Puts Business at Heart of Sustainable Development. January 21. https://www.unilever.com/news/news-and-features/2016/New-global-commission-puts-business-at-heart-of-sustainable-development.html (Accessed October 1, 2016).

United Nations. 2016a. Biodiversity and Ecosystems. https://sustainabledevelopment. un.org/topics/biodiversityandecosystems (Accessed September 20, 2016).

United Nations. 2016b. Goal 6: Ensure Access to Water and Sanitation for All. http://www.un.org/sustainabledevelopment/water-and-sanitation/ (Accessed September 20, 2016).

United Nations. 2016c. Sustainable Development Goals. https://sustainabledevelopment. un.org/?menu=1300 (Accessed September 8, 2016).

United Nations. 2016d. Vast Majority of World—6.76 Billion People—Living with Excessive Air Pollution—UN Report. September 27. http://www.un.org/sustainabledevelopment/blog/2016/09/vast-majority-of-world-6-76-billion-people-living-with-excessive-air-pollution-un-report/ (Accessed October 3, 2016).

US Environmental Protection Agency. 2016. Emergency Response to August 2015 Release from Gold King Mine. October 11. https://www.epa.gov/goldkingmine (Accessed September 27, 2016).

US Green Building Council. 2016. LEED. http://www.usgbc.org/leed (Accessed September 14, 2016).

2

Market Drivers for Greener Products

Consumer Demand for Greener Products

A 2015 global survey conducted by Nielsen found that 66% of respondents are willing to pay more for sustainable goods, up from 55% in 2014. The survey consisted of 30,000 consumers in 60 countries across the world. And it's not just the wealthy that are willing to pay more—results were consistent across regions and income levels. In fact, those earning $20,000 or less annually are actually 5% more willing than those with incomes greater than $50,000 to pay more for products that align with their social and environmental values. This survey demonstrates the growing desire of consumers for more sustainable products.

Of those respondents willing to pay more, a majority were influenced by several key sustainability factors. The top sustainability purchasing drivers include product being made from fresh, natural, and/or organic ingredients (69%), products from a company known for being environmentally friendly (58%), and products from a company known for its commitment to social value (56%). For this group of respondents, personal values clearly outweigh personal benefits, such as cost or convenience (Nielsen 2015b).

Topping the list, however, of sustainability purchasing drivers for a majority of consumers was a single factor: **brand trust**. "Large global consumer-goods' brands that ignore sustainability increase reputational and business risk." This is an opportunity to build brand loyalty with socially conscious consumers looking for products that align with their values. Shelton Group's 2016 Eco-Pulse study backs this up, stating that half of Americans have either started or stopped buying particular brands because of the environmental reputation of the manufacturer, up from just 12% in 2013 (Enkema 2016).

"Environmental issues and concerns about sustainability have become core concerns of society—and must become central elements of business strategy." says Dan Esty Yale Professor and author of Green to Gold. (Esty2011)

Proctor & Gamble is a prime example of consumer values translating into sales. Despite being the world's largest consumer products company, Procter & Gamble is struggling to maintain their stronghold in several product categories as environmentally friendly products surge in popularity. Companies with products that are seen as more natural, environmentally sensitive, or purpose-driven, such as Method, Honest Co., and Seventh Generation, are rapidly chipping away at Procter & Gamble's sales (Very 2016). To maintain their leadership position, Procter & Gamble will have to focus on making many of its own brands greener to appeal to sustainability-minded shoppers.

Retailer's Demands

The greatest driver for developing greener products is when the marketplace demands it. When your customer is asking you for products that have lower environmental impacts, you pay attention. This is especially true when it's your largest customer.

In 2005, Lee Scott, at the time CEO of Walmart Stores, Inc., gave a speech—the first of its kind—to be livecast to all the company's stores. "If we were a country, we would be the 20th largest in the world. If Walmart were a city, we would be the fifth largest in America... What if we used our size and resources to make this country and this earth an even better place for all of us: customers, associates, our children, and generations unborn?" Thus began Walmart's sustainability journey (Makower 2015).

When **Walmart**, the largest retailer in the world, embarked on an aggressive sustainability program and asked their suppliers to help by providing greener products, it changed the way business is conducted. Whatever the reasons are for Walmart's sustainability initiatives, it has been hugely impactful. I cannot think of a more significant single event that propelled the development of greener products than when Walmart decided to embrace sustainability.

When the biggest retailer in the world puts out a sustainability scorecard that can help (or hurt) your sales, you stand up and pay attention.

The company set out a three-step plan: (1) develop a supplier sustainability assessment, (2) develop a life-cycle analysis database, and (3) develop a simple tool that customers can use to consume in a more sustainable way. (Makower 2016)

Walmart Sustainability Goals

- Produce zero waste
- Supplied with 100% renewable energy
- Sell sustainable products

If you sell to Walmart, you will be asked to help provide sustainable products and help them to produce zero waste. Part of the first steps to getting sustainability information on products sold in their stores is through the supplier sustainability assessment. This is a survey consisting of 15 questions that address energy and climate, material efficiency, natural resources, and people and community. Some of the questions asked will no doubt get companies scrambling and spur action where there had been little or none, perhaps in part due to fear that their competitors could get an edge over them. Consider the implications of the following questions; if your answer is no and your competitor has programs in place, this can put you at a competitive disadvantage.

- Have you measured and taken steps to reduce your corporate greenhouse gas emissions? If yes, what are those targets?
- Have you set publicly available water-use reduction targets? If yes, what are those targets?
- Have you obtained third-party certifications for any of the products that you sell to Walmart?
- Do you invest in community development activities in the markets you source from and/or operate within? (Walmart 2016a)

Walmart was not known for sustainability, but over the last 10 years has established itself as a leading force. The company has been extremely successful by several measures. By the end of 2015, Walmart had eliminated 28.2 million metric tons of greenhouse gasses from its supply chain, doubled the fuel efficiency of its fleet from a 2005 baseline, increased its electricity supplied by renewable energy to 26%, and in 2014, diverted 82.4% of its waste from landfills (Makower 2015). Now, more than a decade after Lee Scott set the first sustainability goals for Walmart, the company continues to up their game with a new set of goals for 2025. Over the next 10 years, the company plans to:

- Achieve zero waste to landfill in Canada, Japan, the UK, and the USA.
- Be powered by 50% renewable energy sources under a plan designed to achieve science-based targets
- Double sales of locally grown products
- Expand sustainable sourcing to cover 20 key commodities, including bananas, grapes, coffee, and tea
- Use 100% recyclable packaging for all private-label brands
- Expand sourcing of commodities produced with zero net deforestation (Walmart 2016b)

When you consider the quantity and diversity of products that are sold in their stores around the world, there has been a profound impact on their supply chain. Besides typical consumer brands, there are gardening products, pharmacy products, eye care, home furnishing, electronics, and much more.

All of their suppliers are being forced to consider the sustainability of their products in a way they may have never had to. I can tell you for my company, we are well aware that whenever we have a meeting with Walmart we need to be prepared to discuss what our brands are doing about sustainability.

Tesco is Britain's largest retailer, and one of the top three global retailers. With operations in 13 countries, suppliers in over 70 countries, over 6,900 stores, and over 470,000 employees, this company has significant influence. To fully understand the impact of their operations, Tesco determined that their supply chain in the UK is responsible for approximately 26 million tons of CO_2, which is about ten times the amount from their own operations.

In their commitment to minimize climate change, they set several objectives:

- Become a zero-carbon business by 2050
- Reduce CO_2 emissions in the products in their supply chain, against a 2008 baseline, by 30% by 2020
- Help customers to reduce their carbon footprint by 50% by 2020
- Halve emissions from their 2006/2007 baseline portfolio of buildings by 2020
- New stores built between 2007 and 2020 to emit half the CO_2 of a 2006 new store
- Reduce CO_2 emissions per case of goods delivered against a 2011/2012 baseline by 25% (Tesco 2016)

In addition to climate change, Tesco is committed to forest and marine sustainability. The company has set a goal to achieve zero net deforestation by 2020. In order to do this, they have mapped the supply chains for several products that they deem the biggest global drivers for deforestation (palm oil, cattle products, soy products, and timber), and have put in place sustainable procurement policies. As a result, 99% of Tesco-brand food products are sustainably sourced. Similarly, they have outlined aquaculture requirements that apply to all Tesco-brand aquaculture producers to ensure good farming practices. Tesco also collaborates with the Sustainable Fisheries Partnership to map and risk assess all their seafood supply chains (Tesco 2016). As we can see, many of Tesco's commitments apply to suppliers; if you want to sell in their stores you will have to be mindful of these targets.

Home Improvement Companies

Home improvement companies have gotten in on the greener product movement too, seeing opportunities to align their offerings with their customers' growing desire for products with environmental and social benefits. **Lowe's** has set a sustainability strategy that emphasizes bringing greener products to customers. The strategy is to provide environmentally responsible products, packaging, and services at everyday low prices.

Lowe's Policy on Sustainability

- Educate and engage employees, customers, and others on the importance of conserving resources, reducing waste, and recycling
- Use resources—energy, fuel, water, and materials—more efficiently and responsibly to minimize environmental footprint
- Establish sustainability goals and objectives
- Review and communicate progress made toward achieving established goals and objectives
- Engage on public policy issues related to sustainability (Lowe's 2016)

It is obvious that suppliers offering products with improved performance will get preference in Lowe's stores. Lowe's reports progress based on the environmental benefits from products sold in the form of energy and water savings. In 2015, Lowe's sold enough Energy Star products to:

- Reduce the amount of pollution which was equivalent to taking 2.8 million cars off the road.
- Save consumers more than $2.3 billion each year on their energy bills compared with non-ENERGY STAR-qualified products.

The number of WaterSense-labeled toilets and bathroom faucets sold in 2015 can save enough water in a year to:

- Save more than 2.6 billion gallons of water annually, equivalent to the amount used by 17,500 households
- Save consumers $25.7 million each year on water bills (Lowe's 2015)

Now that's an amazing reduction in environmental impacts and dollars saved through the sale of products!

Their competitor, **Home Depot,** has also set goals to bring greener products to their customers. One objective is to encourage customers to become environmentally conscious shoppers. In 2007, Home Depot launched the at-a-glance Eco Options identification system, which helps educate consumers about products with improved environmental performance.

Home Depot offers over 3,500 Eco Option products. A product is classified as having improved performance if it demonstrates benefits in one of five areas: energy efficiency, water conservation, healthy home, clean air, and sustainable forests. Improvements are judged by third-party certifications that have been given to products like the U.S. EPA's Design for Environment, USDA Organic, Forest Stewardship Council, U.S. EPA Energy Star, and other criteria (Home Depot 2016).

The type of products that are given the Eco Options designation include low VOC paint, Water Sense®-labeled bathroom fixtures, Energy Star®-labeled electrical products, organic plant food, and environmentally preferred cleaners. Any company wishing to sell products in a Home Depot store will need to pay attention to their Eco Options program and seek this designation. In addition to the improved environmental performance, these greener products give customers cost savings. This makes them even more attractive and encourages manufacturers to develop more sustainable products.

Greener Hospitals

Companies providing products to hospitals are not exempt from the greener product revolution. Why would hospitals be demanding the development of greener products? Consider that they are operating 24/7. The lights are always on, there is waste being constantly generated, disinfectants and various other chemicals are being used, air pollutants are generated by boilers that supply heat and hot water, and wastewater is continually flowing. There are financial benefits for reducing a hospital's footprint—according to Kathy Gerwig, VP and Environmental Stewardship Officer for Kaiser Permanente, their Environmentally Preferred Purchasing program resulted in $63 million in annual savings from reducing energy, waste, and toxic chemicals (Gerwig 2015).

Hospitals are being encouraged to become more sustainable by interest groups like Health Care Without Harm. Using a phrase from the Hippocratic Oath that doctors take, their mission is to see health care **"first, do no harm."** Their goal is to encourage health care providers to do away with practices that harm people and the environment. The link between human health and environmental pollution is a point used to enroll more hospitals.

Product manufacturers are impacted by this movement since there is a focus on purchase of safer products, materials, and chemicals. Hospitals are trying to avoid products containing toxic materials such as mercury, polyvinyl chloride (PVC) plastic, and brominated flame retardants (Health Care Without Harm 2016).

One of the leading health care providers, **Kaiser Permanente,** raised the bar for greener health care products when it unveiled its Sustainability Scorecard in 2010. Each company intending to sell to Kaiser Permanente is to complete the scorecard and the results will be used to make purchase decisions. I have heard it stated at public conferences that a product's sustainability performance can be up to 20% of the purchase decision depending on product category. So companies need to take their scorecard seriously if they want to sell to Kaiser.

One of the focus areas is the use of toxic substances; an example would be understanding if a product contains di (2-ethylhexyl) phthalate, or DEHP. This chemical is undesirable because it has shown adverse effects on the development of the male reproductive system in young laboratory animals, and there is some concern this could also affect some human patients. So, one question on Kaiser Permanente's scorecard is "Does the product contain DEHP?"

If the answer is "no," the vendor enters a 0; if the answer is "yes," it scores 1. The lower-scored products are the more environmentally friendly. The implementation of this innovative scorecard can impact how medical-device manufacturers do business since Kaiser Permanente purchases more than $1 billion each year of medical products (Hicks 2010).

In June 2016, Kaiser Permanente announced their new "Green Goals" for 2025. These targets include:

- Reduce water by 25% per square foot
- Recycle, reuse, and compost 100% of non-hazardous waste
- Become "carbon net positive" by buying enough clean energy and carbon offsets to remove more greenhouse gases from the atmosphere than it emits
- Buy all of its food locally or from farms and producers that use sustainable practices, including the use of antibiotics responsibly
- Increase its purchase of products and materials that meet environmental standards to 50%
- Meet the ISO 14001 international standards for environmental management at all its hospitals
- Pursue new collaborations to reduce environmental risks to the foodsheds, watersheds, and air basins supplying its communities (HCO News 2016).

To achieve these goals, Kaiser Permanente will have to rely heavily on their suppliers providing greener products. With their large purchasing power, the health care industry is an important driver of more sustainable products.

B2B Purchasing

Green products are not only relevant to consumers, we have also seen a strong pull from business-to-business (B2B) customers. The phrase, "greening the supply chain" has become synonymous with one business asking another to become more sustainable. Companies are pressured to become more sustainable on many fronts. One area that was not originally foreseen was the focus on the supply chain, from procurement of services and raw materials to third-party manufacturers.

More companies are asking their suppliers to help them with their sustainability goals. Unilever has been pressured for using palm oil and other agricultural raw materials in their products that are sourced from farms that have damaged tropical rain forests. To address this issue, they have been working with suppliers to commit to 100% sustainably sourced palm oil, tea, soy beans, and other agricultural products (Unilever 2016). Walmart set goals for its suppliers in China to reduce packaging and increase energy efficiency of products sold in their stores (Walmart 2014). Staples, the office supply company, has implemented a Sustainable Paper Procurement Policy, which ensures that all paper products they sell are sourced in an environmentally and socially responsible manner. Through these initiatives, they are improving the forest-management practices of their suppliers and protecting endangered and high-conservation-value forests (Staples 2016). SC Johnson, in their efforts to minimize impact on the environment and support universal human rights, has developed the SC Johnson Supplier Code of Conduct, which specifies the minimum requirements that all SC Johnson suppliers are required to comply with (SC Johnson 2016).

Procter & Gamble (P&G) has also developed a supplier Environmental Sustainability Scorecard. The scores that suppliers receive will be used to assign an overall rating from P&G which is used to make determinations on who they give their business to. One of the goals of the scorecard is to encourage suppliers to implement more sustainability initiatives. Suppliers must provide data on electrical and fuel use, water input and output, Scope 1 and Scope 2 greenhouse gas emissions, waste sent to landfill or incinerated, and hazardous waste disposal (Procter & Gamble 2016). P&G believes this rating system can encourage environmental improvement across its entire supply chain. The impact can be huge since they have approximately 75,000 suppliers resulting in about $65 billion in annual spending (Procter & Gamble 2016). This scorecard will surely give firms with greener attributers to their product an edge over others.

These are just a few examples of how businesses are looking to their suppliers to help with their sustainability initiatives and to help them green up their products. Companies are no longer only being asked to be responsible for their own footprint but are being held accountable for addressing their entire supply chain. Suppliers with a good sustainability story can gain an edge with their key customers.

Eco-Innovation as a Value Driver

More companies are seeing sustainability as a way to drive innovation to generate new products. There are several ways sustainability can add value to your brand.

Communicating product attributes that emphasize a brand's sustainability is an easy way to get some quick wins.

Addressing a growing apprehension to use bottled water, Brita water filters (a Clorox product) initiated a repositioning of its product.

Brita took advantage of consumers' growing desire to become greener by highlighting the use of their product as more sustainable than bottled water. They calculated the millions of empty bottles that would be taken out of the waste stream to make the case that their product is the more sustainable option for getting clean drinking water. Every one Brita filter used saves over 300 standard 16.9 oz. water bottles from being used and discarded. On their website you can see an estimated amount of bottle use reduced through use of the Brita filter system (over 430 million fewer estimated bottles used as of 2016). The overall product positioning is all about how a customer can help the environment by using their product.

If you go to the Brita website you will see this very clearly, "better for the environment and your wallet." In other words, you save the environment and money at the same time! A single Brita pitcher with filters can save consumers almost $1000 over buying standard water bottles (Brita 2016). The Brita water filter repositioning is an excellent example of using innovative thinking to uncover existing sustainability attributes of your product without having to do any physical changes to the product itself.

Another example of repositioning products by highlighting their existing greener attributes is GE's Ecomagination. Under the Ecomagination banner, GE can position any product that has an environmental benefit as a more sustainable option to their customers. GE's CEO Jeff Immelt, realized that they were selling windmills, had more efficient locomotive and jet engines, and were recognized as a US EPA Energy Star partner, which sparked his epiphany: "Maybe there is something there if we put all of those together" (Makower 2009). Consider washing machines that use less water than competitors (or previous versions), hot water heaters and microwave ovens that use less energy—all of these can be positioned under the Ecomagination banner. The reason why the equipment has better environmental performance doesn't matter; the key is *if it does*, communicate it to your customers!

Use of sustainability as the driver for developing new product concepts is another way to meet market needs.

A good example of this is Samsung's greener product development process and eco-logo called PlanetFirst™. This process is used to generate innovative new products that have substantial sustainability benefits. Examples of these new innovative products include an LED TV (UE60J6150) that uses less power in normal operation and in standby mode, with an auto power-down function and a light-intensity sensor, resulting in significant operational cost savings for the customer. The Galaxy S6 Edge mobile phone has a high-efficiency charger (82%), ultra power-saving mode, and uses 100% recycled paper in its packaging (Samsung 2016).

Another example of sustainability as an innovation driver is The Clorox Company's line of natural cleaners, Green Works®. Clorox's cleaning and disinfecting products are not particularly known for being environmentally friendly. However, the company made a fresh start using sustainability as the basis and came up with a new brand of cleaning products called Green Works®. These household cleaning products are made from high-quality ingredients that are at least 97% naturally derived (The Clorox Company 2016).

It is common for companies to run innovation sessions to develop new product concepts. As we have seen, many companies have been successful in using sustainability as the innovation driving force. I have had the opportunity to run some **eco-innovation sessions** and have been surprised to see that top executives are unfamiliar with the sustainability achievements that their business unit has achieved. If the executives are unaware, our customers are as well. With the growing demands for more sustainable products in all types of businesses, it is a wasted opportunity to not communicate the improvements made already and to generate new product concepts.

Running an Eco-Innovation Session

Eco-innovation doesn't have to be a complicated thing. Simple tools can be used to spur innovative thinking. Some of the key groups that you would want to include in an innovation session would be marketing, sales, R&D, operations, procurement, environment, health & safety, and communications. A simple agenda for one of these events would look like this.

Eco-Innovation Session Agenda

- *Landscape*—What's going on in the marketplace? Are customers looking for greener brands?
- *Competitor Analysis*—Is anyone leading in sustainability? Are we competing against any green brands? Do any competitors have weaknesses compared to our brand?
- *Company Accomplishments*—What have we accomplished as an enterprise and are there any specific greener attributes to our brand (e.g., use of recycled content, better performance, end-of-life solutions)?

- *Risk Analysis*—What happens if we do nothing? Do we have any sustainability problems (perceived or actual)?
- *Opportunity Analysis*—(1) What can we communicate to customers that we have accomplished already? and (2) what can we do to green up our brand or develop a new product?
- *Prioritization*—Develop a prioritized list of projects and assignments (Unruh 2010).
- If developing a new product concept or greening up an existing brand is a desired outcome of the innovation meeting then the Eco-design elements in Figure 2.1 are a good way to initiate innovative thinking.

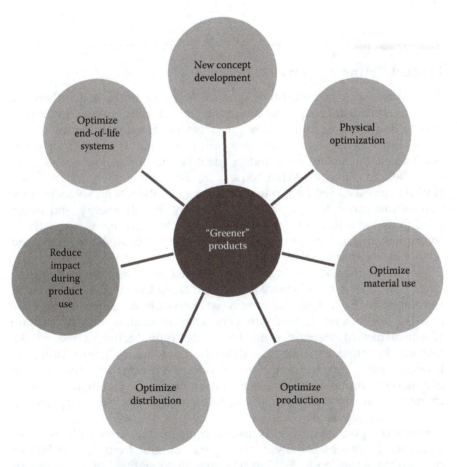

FIGURE 2.1
Eco-design elements. (From Brezet and Hemel, PROMISE Manual of Ecodesign, adapted by Natural Resource Council Canada, 2007, Design for Environment Guide. Five Winds International. Ottawa, Canada, p. 19.)

Just taking a quick look at Figure 2.1 can spark many sustainable product innovations. If customers have been complaining about excessive packaging or that it can't be recycled, then we should initiate a project to reduce the packaging size and use commonly recycled materials. Once the improvement is accomplished, make sure you tell them about it. The engineering department mentioned that we reduced the energy use of the product by 10% in its latest design and we think we are best in class for energy use—we should do a competitor analysis and let our customer know that our product is saving them money and is better for the environment. We are losing some share points to a competitor with a new natural brand; let's develop our own natural product line or try to "green up" by removing ingredients that the marketplace is concerned with.

Product Rating Systems

Another reason to be mindful of a product's sustainability is that there are many independent groups that are rating your product and making it very easy to find out the score. "There's no room for green-washing anymore," states Chris Sayner from specialty chemicals company Croda. "People just won't accept your word for it: independent verification is now necessary to give true transparency" (Whitehouse 2016). Environmental Working Group (EWG) is an NGO with a product-rating organization, with six major program areas: toxics, food, agriculture, children's health, energy, and water. EWG's mission is threefold: **advocate** for a healthier, greener world, **activate** consumers to stand up for better policies and markets, and **advance** sound science in environmental health.

The Environmental Working Group has developed several product ratings. One of the most comprehensive is the **Skin Deep Cosmetic rating system.** Developed in 2004, there are over 80,000 products rated. The ratings are based on two factors: a hazard rating and a data availability rating. There are 17 general hazard categories evaluated: cancer, reproductive/developmental toxicity, neurotoxicity, endocrine disruption potential, allergies/immunotoxicity, restrictions/warnings, organ system toxicity, persistence/bioaccumulation, multiple/additive exposure, mutations, cellular/biochemical changes, ecotoxicity, occupational hazards, irritation, absorption, impurities, and miscellaneous.

A very complicated method is used to come up with an overall rating for a product. There are 260 individual categories ranging from "known human carcinogens according to EPA" to "skin irritants identified by the Cosmetic Ingredient Review panel." The categories are then mapped into one of 205 "score categories." All scores are scaled from 0 to 10, with 10 being the highest level of concern and 0 the least.

Some of their worst-rated products can have some scary sounding information about them. As an example, when evaluating the worst-rated lip gloss, it received a score 9 (10 being worst)—an overall high hazard listing for ingredients—and had the following information listed in the product details (EWG 2016).

Ingredients in the product are linked to:

- Cancer
- Developmental/reproductive toxicity
- Allergies/immunotoxicity

Other concerns for ingredients used in this product:

Biochemical or cellular level changes, endocrine disruption, additive exposure sources, irritation (skin, eyes, or lungs), occupational hazards, persistence and bioaccumulation; other moderate concerns: cancer, contamination concerns, organ system toxicity (nonreproductive) (EWG 2016).

Other than that, it is a pretty good product! Any customer or potential customer that happens to review this database would, at the very least, question their purchase of this product.

In addition to the ratings they provide, EWG has launched their **EWG VERIFIED**™ program, which highlights products that go above and beyond its green rating in EWG's Skin Deep because the company has disclosed more about its formulations and manufacturing processes. To become EWG VERIFIED™, a product must meet all of EWG's strictest criteria, including:

- Products must score a "green" in EWG's Skin Deep database
- Products cannot contain any ingredients on EWG's "unacceptable" list, meaning ingredients with health, eco-toxicity and/or contamination concerns
- Products cannot contain any ingredients on EWG's "restricted" list, which do not meet the restrictions set by authoritative bodies and industry institutions
- Products must follow standard ingredient naming guidelines
- Products must fully disclose all ingredients on the label, including ingredients used in fragrance
- Product manufacturers must develop and follow current good manufacturing practices
- Products must follow the European Union requirements for labeling fragrance allergens

- Products must follow the European Union labeling guidelines for nanomaterials used in cosmetics
- Product labels must indicate an expiration date or a "period of time after opening" (EWG 2016)

The EWG VERIFIED program makes it easy for consumers to identify the highest-rated products. Retailers have created similar labels for their most sustainable products. Target has introduced "Made to Matter," a collection of products from purpose-driven brands that make natural, organic, and sustainable products more accessible. Walmart has a "Sustainability Leaders" badge that identifies over 10,000 items made by companies identified as leaders in a category (like televisions or plastic toys) through their Sustainability Index.

Regardless of how valid a company may feel these rating or labeling systems are, it merits to pay attention to them and determine whether there is something to be learned that could influence product improvements. At the very least, it gives a perspective that is not always intuitive to the typical manufacturer.

Socially Responsible Investment

Another trend that is encouraging the development of sustainable products is socially responsible investing. More investors are considering issues beyond corporate profit, and are interested in purchasing stock only from companies that align with their personal values. According to Peter Roselle, vice president of Morgan Stanley, socially responsible investing can be done without sacrificing returns on investment. In fact, research shows it may actually increase returns! (Roselle 2016). Rating systems such as the Dow Jones Sustainability Indexes (DJSI) and the FTSE4Good indexes are prominent metrics that encourage socially responsible investing.

The DJSI track the performance of companies that lead the field in terms of corporate sustainability (DJSI 2016). The FTSE4Good Index Series, based in the United Kingdom, is designed to measure the performance of companies that meet globally recognized corporate responsibility standards, and to facilitate investment in those companies (FTSE 2016).

In order to fare well in these indexes, a company must be performing well in all aspects of sustainability, including bringing greener products to market. More focus is being placed on product stewardship, eco-efficiency, and supply-chain issues associated with product manufacturing. Therefore, failure to adequately address product stewardship issues like potentially toxic materials in your product, not adequately addressing product end-of-life issues or failing to initiate sustainability programs at your suppliers, can hurt your rating in these indexes.

Shareholder Resolutions

In addition to social responsibility funds and investing pressures, we have seen an increasing amount of **shareholder resolutions** focused on product responsibility. Shareholder proposals focusing on environmental, social, and policy issues are at an all-time high, accounting for half of all proposals at the largest 250 US companies in 2016, overtaking proxy access as the main target of shareholder activism (Stein 2016). Using shareholder resolutions is a really smart strategy which environmental groups have initiated. As long as they own a share in a publicly traded company, they can propose a proxy statement which shareholders would have to vote on according to the U.S. Securities and Exchange Commission requirements. Corporations go through great lengths to try and address the concerns of environmental groups to prevent these resolutions from going to a shareholder vote to manage public perception of their company.

An example of the type of product-related questions that corporations are being asked would be a resolution calling on Duke Energy, the biggest carbon pollution emitter of any US power producer, to publish a report assessing the public health impact of its coal use on rates of illness, mortality, and infant death, due to coal related air and water pollution in communities adjacent to Duke's coal operations.

Similarly, BP and Shell will be addressing resolutions seeking greater disclosure about climate change. They are expected to issue reports regarding fossil fuels and their contribution to climate change (CBIS 2015).

Fast food giants such as McDonald's, KFC, and Chick-fil-A are being asked to stop routine antibiotic use on their chicken. The World Health Organization warns that the overuse of antibiotics may eventually lead to many infections becoming untreatable due to antibiotic resistance. These resolutions are getting a lot of attention since 70% of antibiotics sold in the United States are for livestock and poultry use (Fortune 2016).

Common Shareholder Product Stewardship Resolutions

- Sustainable sourcing of raw materials
- Addressing climate change
- Use of GMO materials
- Reducing toxic chemical usage

Green Public Procurement

A significant new approach to procurement is the purchase of goods and services that foster lower environmental impact. This Green Public Procurement (GPP) is a process whereby public authorities seek to procure goods, services, and works with a reduced environmental impact throughout their life cycle.

In Europe they have realized that green purchasing can have a big impact. The European Union public authorities (central, regional, and local levels) spend approximately 14% of EU GDP—or €1.8 trillion—on goods, services, and works each year. A good deal of this spending is on services that have significant environmental impact, such as transportation, buildings, and food. By using GPP criteria, public officials can sway the purchase of items to reduce their impact (European Commission 2016).

European Union member states have developed National Action Plans (NAPs) for greening their public procurement. The NAPs contain targets that are reported on publicly. The type of goods and services that are of primary focus include paper, cleaning chemicals, office IT equipment, construction, transport, furniture, electricity, windows, doors, thermal insulation, road construction, and mobile phones (European Commission 2016). Any company that manufactures products in these categories should be interested in evaluating the criteria that the EU has set for GPP.

In the United States, the Environmental Protection Agency has been tasked with the development of Environmentally Preferable Purchasing (EPP) guidance for federal agencies to implement. Similar to the initiative in Europe, the impact can be significant. "The Federal government is the single largest purchaser in the United States, spending over $350 billion each year on a wide variety of products and services." This purchase power carries with it a significant environmental footprint.

The Federal government can minimize environmental impact through the purchase of goods and services by trying to purchase those with lowest impacts. The EPA defined EPP as products or services that "have a lesser or reduced effect on human health and the environment when compared with competing products or services that serve the same purpose. This comparison may consider raw materials acquisition, production, manufacturing, packaging, distribution, reuse, operation, maintenance or disposal of the product or service." Guiding principles were developed to help federal agencies follow through with these commitments.

US EPA Guiding Principles for Environmentally Preferred Purchasing

1. Environment + Price + Performance = EPP
   Include environmental considerations as part of the normal purchasing process.

2. Pollution Prevention
   Emphasize pollution prevention as part of the purchasing process.
3. Life-Cycle Perspective/Multiple Attributes
   Examine multiple environmental attributes throughout the product and service's life cycle.
4. Comparison of Environmental Impacts
   Compare environmental impacts when selecting products and services.
5. Environmental Performance Information
   Collect accurate and meaningful environmental information about environmental performance of products and services.

(USEPA 2016)

Scorecards were developed to help federal agencies to track the progress they were making, particularly the President's Office of Management and Budget. Here again, companies that sell to the federal government need to be mindful of the potential impacts of this initiative. Consider losing a sale to a large customer like the US government because your product was not on par with your competitors' product.

Conclusion

The market is demanding greener products and this demand is growing. Manufacturers need to fully realize the growing pull for products with enhanced environmental and social benefits.

Market Demands for Greener Products

- Consumers desire greener products
- Institutional customers are requesting them
- Hospital green revolution
- B2B green purchasing
- Eco-innovation as a value driver

No matter what aspect of business you are in, there is a shift occurring that makes it imperative to offer greener products to your customers. We are seeing market demands in all major sectors: consumer goods, to chemicals, transportation, medical products, pharmaceuticals, energy, and others. Aside from customer demand, there is pressure from other areas such as environmental groups, sustainability rating organizations, and competitors. Sustainability is even being used as an innovation driver for new product development. Having greener product offerings is no longer "a nice to have" prospect, but a necessity in order to be competitive in the marketplace.

References

Brita. 2016. Why Brita. https://www.brita.com/why-brita/#world (Accessed November 2, 2016).

CBIS. 2015. BP joins Shell in Landmark Support for "Aiming for A" shareholder resolutions co-filed by CBIS. http://cbisonline.com/us/news/shells-landmark-decision-support-aiming-shareholder-resolution-co-filed-cbis/ (Accessed December 18, 2016).

DJSI. 2016. Dow Jones Sustainability Indexes. http://www.sustainability-index.com/ (Accessed November 28, 2016).

Enkema, Susannah. 2016. Changing Attitudes to Habits: How Do We Move People from Green Attitudes to Green Habits? Shelton Group. September 22. http://sheltongrp.com/changing-green-attitudes-to-lasting-sustainable-habits/ (Accessed October 26, 2016).

Environmental Working Group (EWG). 2016. Skin Deep Guide to Cosmetics. http://www.ewg.org/ (Accessed November 20, 2016).

European Commission. 2016. Environment FAQs. http://ec.europa.eu/environment/gpp/faq_en.htm (Accessed November 28, 2016).

Fortune. 2016. Consumer Groups Push KFC to Stop Routine Antibiotic Use in Its Chicken. August 10. Retrieved from http://www.asyousow.org/wp-content/uploads/2016/08/20160810-fortune-Consumer-Groups-Push-KFC-to-Stop-Routine-Antibiotic-Use-in-Its-Chicken.pdf (Accessed December 18, 2016).

FTSE. 2016. FTSE4Good Index Series. http://www.ftse.com/Indices/FTSE4Good_Index_Series/index.jsp (Accessed November 28, 2016).

Gerwig, Kathy. 2015. *Greening Health Care.* Oxford University Press, New York, p. 178.

HCO News. 2016. Kaiser Permanente Makes Aggressive Green Goals for 2025. June 7. http://www.hconews.com/articles/2016/06/7/kaiser-permanente-makes-aggressive-green-goals-2025 (Accessed November 2, 2016).

Health Care Without Harm. 2016. Mission and Goals. http://www.noharm.org/all_regions/about/mission.php (Accessed November 2, 2016).

Hicks, Jennifer. 2010. Sustainable Scorecard to Be Used to Help Kaiser Permanente Evaluate Medical Products. May 4. http://www.triplepundit.com/2010/05/kaiser-permanente-sustainable-scorecard/ (Accessed November 2, 2016).

Home Depot. 2016. Eco Options. http://ecooptions.homedepot.com/ (Accessed November 4, 2016).

Lowe's. 2015. 2015 Social Responsibility Report. https://1g0r7s45brd833po5f1d5yyb-wpengine.netdna-ssl.com/wp-content/uploads/2016/04/lowes-2015-social-responsibility.pdf (Accessed November 4, 2016).

Lowe's. 2016. Lowe's Policy on Sustainability. https://www.lowes.com/cd_lowes policyonsustainability_1286385507_ (Accessed November 4, 2016).

Makower, Joel. 2009. The Rise of Ratings. In J. Makower, Strategies for the Green Economy (p. 84). McGraw Hill, New York.

Makower, Joel. 2015. Walmart Sustainability at 10: The Birth of a Notion. GreenBiz. November 16. https://www.greenbiz.com/article/walmart-sustainability-10-birth-notion (Accessed November 2, 2016).

Makower, Joel. 2016. *Walmart Sustainability at 10: An Assessment.* November 17. https://www.greenbiz.com/article/walmart-sustainability-10-assessment (Accessed November 4, 2016).

Nielsen. 2015a. *Consumer-Goods' Brands That Demonstrate Commitment to Sustainability Outperform Those That Don't.* October 12. http://www.nielsen.com/us/en/press-room/2015/consumer-goods-brands-that-demonstrate-commitment-to-sustainability-outperform.html (Accessed October 26, 2015).

Nielsen. 2015b. *The Sustainability Imperative.* October 12. http://www.nielsen.com/us/en/insights/reports/2015/the-sustainability-imperative.html (Accessed October 26, 2016).

Proctor & Gamble. 2016. *2016 Annual Report.* http://www.pginvestor.com/Cache/1500090608.PDF?O=PDF&T=&Y=&D=&FID=1500090608&iid=4004124 (Accessed November 2, 2016).

Proctor & Gamble. *Supplier Efforts.* http://us.pg.com/sustainability/environmental-sustainability/supplier-efforts (Accessed November 2, 2016).

Roselle, Peter. 2016. The Evolution of Integrating ESG Analysis into Wealth Management Decisions. *Journal of Applied Corporate Finance* 28, 75–79.

Samsung. 2016. *Environment.* http://www.samsung.com/us/aboutsamsung/sustainability/sustainabilityreports/download/2016/samsung-sustainability-report-2016-environment.pdf (Accessed November 12, 2016).

SC Johnson. 2016. *Supplier Code of Conduct.* http://www.scjohnson.com/en/commitment/supplychaintransparency/suppliercodeofconduct.aspx (Accessed November 2, 2016).

Staples. 2016. *Sustainable Products.* http://www.staples.com/sbd/cre/marketing/about_us/sustainable-products.html#z_greener_paper (Accessed November 2, 2016).

Stein, Mara L. 2016. *The Morning Risk Report: In 2016 Proxy Ballots, Access is Out, Environmentalism is In.* Septmeber 7. http://blogs.wsj.com/riskandcompliance/2016/09/27/the-morning-risk-report-in-2016-proxy-ballots-access-is-out-environment-is-in/ (Accessed November 28, 2016).

Tesco PLC. 2016. *Tesco and Society.* https://www.tescoplc.com/tesco-and-society/sourcing-great-products/reducing-our-impact-on-the-environment/ (Accessed November 4, 2016).

The Clorox Company. 2016. *Green Works.* https://www.greenworkscleaners.com/ (Accessed November 12, 2016).

Unilever. 2016. *Transforming the Palm Oil Industry.* https://www.unilever.com/sustainable-living/the-sustainable-living-plan/reducing-environmental-impact/sustainable-sourcing/transforming-the-palm-oil-industry/ (Accessed November 2, 2016).

Unruh, Gregory and Ettenson, Richard. 2010. Growing Green. *Harvard Business Review.* 6.

Very, Sarah. 2016. *P&G Under Pressure as Eco-Friendly Products Surge.* October 4. *Chicago Tribune.* http://www.chicagotribune.com/business/ct-proctor-gamble-sustainable-products-20161004-story.html (Accessed November 17, 2016).

Walmart. 2014. *Walmart Continues to Strengthen Global Supply Chain Sustainability; Announces New Commitment to Advance Factory Energy Efficiency in China.* August 27. http://corporate.walmart.com/_news_/news-archive/2014/08/27/walmart-continues-to-strengthen-global-supply-chain-sustainability-announces-new-commitment-to-advance-factory-energy-efficiency-in-china (Accessed November 2, 2016).

Walmart. 2016a. *Supplier Sustainability Assessment: 15 Questions for Suppliers.* www. Walmartstores.com (Accessed November 2, 2016).

Walmart. 2016b. *Walmart Offers New Vision for the Company's Role in Society.* November 4 http://news.walmart.com/2016/11/04/walmart-offers-new-vision-for-the-companys-role-in-society (Accessed November 26, 2016).

Whitehouse, Lucy. 2016. *"No Room for Greenwashing": The Reality of Sustainability for the Personal Care Industry. Cosmetics.* December 15. http://www.cosmeticsdesign-europe.com/Regulation-Safety/Sustainability-and-palm-oil-no-room-for-green-washing (Accessed December 25, 2016).

3

Regulatory Drivers for Greener Products

A New Set of Rules

It used to be that the only environmental regulations a company had to be concerned with were those affecting a manufacturing facility's air emissions, waste generation, and wastewater. Governments began to realize that the disposal of products presented significant environmental concerns and soon began developing regulations to address this issue. There has been an exponential growth of environmental regulations that apply to products over recent years. Beginning in Europe with requirements for developing more sustainable packaging and mandatory take-back requirements, these regulations have expanded to all regions of the world.

Having to comply with this new type of regulation requires manufacturers to develop management systems to insure that products being brought to market comply with the myriad of design, reporting, labeling, and fee requirements throughout the world. Organizations within a company that have not typically had to be concerned with environmental regulations now are faced with new challenges. For example, R&D groups have to develop processes to insure that banned or restricted materials are not in their new products. Sales and marketing units must insure that labeling, registration, and fees are paid to governments where their products are sold. Further, systems have to be set up to facilitate the recovery and recycling of such things as electrical products, packaging, sharps, and unused medications.

Product-based regulations are becoming drivers for new product design and are causing product development teams to anticipate and monitor these requirements. As a case study, consider having to comply with new requirements that restrict the limit of certain toxic metals, flame retardants, and phthalates in electronics. The European Union Restriction of the Use of Certain Hazardous Substances in Electrical and Electronic Equipment Regulations (the "RoHS Regulations") has expanded into other categories beyond consumer electronics such as medical equipment. Addressing these requirements affects product design criteria.

Knowing that regulatory deadlines are approaching, design groups need to start discussions with their suppliers to begin the testing and validation of new RoHS compliant parts. In order to prevent barriers to the sale of your product, designs have to be changed years prior to the compliance date. Companies that can anticipate these regulations and make changes more quickly than the competition can gain in the marketplace by appealing to purchasers that desire greener products, if they can certify compliance ahead of schedule.

Packaging Regulations

Another area of regulation that has complicated product sales is packaging regulations. It is getting increasingly difficult to insure that product packaging is compliant in all regions of the world. As more companies move towards global brands, a single package is being used for all markets; therefore, the design must incorporate a multitude of regulatory requirements. Consider that there are environmental packaging regulations of some type in all regions of the world, and new regulations are being added on a regular basis.

Regions with Environmental Packaging Requirements

- Americas—Brazil, Canada, United States, Mexico
- Africa/Middle East—Israel, South Africa, Tunisia
- Europe—European Union Packaging Directive, Bulgaria, Croatia, Iceland, Norway, Romania, Switzerland, Turkey, Ukraine
- Asia Pacific—Australian, Bangladesh, China, India, South Korea, Japan, Taiwan (EPI 2016)

Let us consider some of the requirements necessary for a package to be sold globally. As an illustration, let's evaluate the standards of one packaging regulation, the EU Packaging Directive.

The EU Packaging and Packaging Waste Directive *(94/62/EC)* requires:

- *Source reduction:* Companies must demonstrate that they have reduced their packaging as much as possible and then identify the critical area (such as product protection, safety, consumer, and acceptance) which prevents further reduction in weight or volume of a packaging component.

- *Recovery standards:* Packaging components must be recoverable by at least one of three recovery routes (energy, organic, or material recovery) and must meet certain requirements specific to that recovery route.

- *Reuse:* Optional, but a package must meet the requirements of the reuse standard if it is claimed as reusable.
- *Heavy metals content:* Sets a concentration limit for lead, cadmium, mercury, and hexavalent chromium in packaging.
- *Reduction of hazardous substances in packaging:* Substances classified as noxious (e.g., zinc) must be minimized if they could be released in emissions, ash, or leachate when packaging is land filled or burned (ODEQ 2005).

In addition to these requirements, fees must be paid based on the type of packaging put on the market. This affects the package design also because there are higher fees for packaging that are not easily recyclable. So to minimize fees, you want to use the least costly (most recyclable) materials in your design. Regulations covering each of these requirements must be developed and put into law for every member state of the European Union. Therefore, each state can have slightly different requirements to meet the objectives. This makes it more complex for companies placing products on the market in the EU (EUR-Lex 2014b).

Similar to the European Union requirement, the government of Korea also has set stringent regulations.

South Korea Packaging Requirements

- Fees for certain packaging
- Restrictions on heavy metals, PVC, and expanded polystyrene
- Labeling of recyclable materials
- Empty space and layering requirements
- Reusable containers encouraged for some products (ODEQ 2005)

Having to conform to different regulations in many countries makes it very problematic to design a globally compliant package. This is just one environmental product regulatory requirement. When you view Figure 3.1 developed by the consulting firm Arcadis and consider all the global regulations that firms must comply with when putting a product on the market, it seems quite daunting.

Restriction on the Use of Chemicals and Notifications

One of the most significant chemical regulations that has brought substantial changes to the sale of products in Europe and throughout the world is the **European REACH** regulation. The acronym REACH stands for

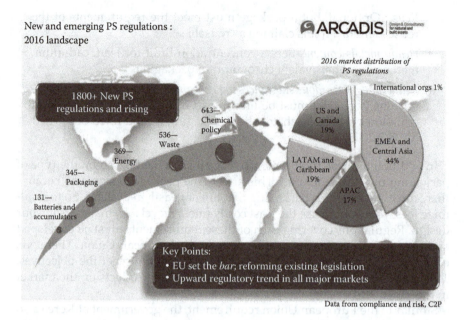

New and emerging PS regulations :
2016 landscape

FIGURE 3.1
New and emerging environmental product regulations. (Courtesy of Arcadis, Amsterdam, the Netherlands.)

Registration, Evaluation, and Authorization of Chemicals. The regulation has been in effect since 2007; it requires that all products and chemicals imported and manufactured in EU member states greater than 1 ton per year be registered with the European Chemicals Agency. Its aim is to place the responsibility on chemical manufacturers and importers to insure chemicals are properly tested and are being used in a way that is protective of human health and the environment. The regulation requires more data on the hazards of chemicals developed, and it restricts or bans "substances of very high concern" (SVHC) (European Commission 2016). This has been a gamechanger for the use of chemicals in products and manufacturing processes and its impact is being seen throughout the world. Besides requiring registrations and much more information on the toxicity of chemicals, companies are facing the prospect of having to find other materials for their products and production processes if they use any chemicals on the SVHC list. Following the example set in the EU, other countries are adopting REACH-like regulations.

For instance, China and Korea REACH have mimicked certain elements of the European REACH requirements. These regulations are similar to the EU REACH but have their own country requirements. The Measures for Environmental Administration of New Chemical Substances (China MEP, Order 7) also known as **China REACH** applies to new chemical substances regardless of the quantity. Any new chemical that is not on the Existing Chemical Substances Produced or Imported in China (IECSC) list (about 45,000 substances) must meet certain notification and testing requirements. Some of the requirements of this regulation include:

- Completion of a notification application form
- A test report detailing the substance's physiochemical properties and its toxicity and eco-toxicity
- An environmental risk assessment report
- Recommended classification and labeling
- Preparation of a Chinese Safety Data Sheets (CIRS 2014)

These REACH-like regulations are becoming more prevalent and are influencing regulatory developments in other regions besides the Asia Pacific. A case in point is the California Green Chemistry Initiative, which looked to REACH for inspiration.

There are many other regulations that restrict the use of chemicals besides REACH. Several of them focus on electronics and electrical equipment because of the hazardous materials employed in their design. Examples include the European Union RoSH, China, Japan, and Korea RoSH requirements. These regulations put very tight limits on the amount of certain compounds that cause harm to human health and the environment if improperly disposed of. As an example, the EU Directive 2002/95 known as RoHS (restriction of certain hazardous substances) bans the placement of new electrical and electronic equipment containing greater-than-specified levels of lead, cadmium, mercury, hexavalent chromium, polybrominated biphenyl (PBB) and polybrominated diphenyl ether (PBDE) flame retardants, and phthalates on the EU market. The limits range from 0.01% to 0.1% depending on the compound (DBIS 2014).

RoHS restricts the levels for the following materials:

RoHS List of Restricted Substances	
Substance Name	Limit (%)
Lead	0.1
Mercury	0.1
Cadmium	0.01
Hexavalent chromium	0.1
Polybrominated biphenyls (PBB)	0.1
Polybrominated diphenyl ethers (PBDE)	0.1
Bis(2-ethylhexyl) phthalate (DEHP)	0.1
Butyl benzyl phthalate (BBP)	0.1
Dibutyl phthalate (DBP)	0.1
Diisobutyl phthalate (DIBP)	0.1

Registration and Restriction of Chemicals in Products

There is also an expanding list of regulations that require products containing certain hazardous chemicals to be registered in the country where the product is sold. The following are examples of regulations that require products to be registered, restricted, or that ban the use of certain chemicals or require labeling of products that contain specific chemicals.

- *EU REACH*: In effect since 2007; affects all products and chemicals imported into and manufactured in EU member states; requirements include registration, communication, and restrictions of SVHC.
- *China REACH*: Similar to EU REACH; requires registration, testing, labeling, and data sheets for new chemicals before they can be put on the market in China.
- *EU RoHS*: Restricts 10 hazardous chemicals used in electrical and electronic equipment (lead, cadmium, mercury, hexavalent chromium, polybrominated biphenyl [PBB] and polybrominated diphenyl ether [PBDE] flame retardants), and four phthalates have been in place since 2006.
- *China RoHS*: Restricts the use of the same EU RoHS chemicals not including the phthalates; effective since 2007; covered products include medical equipment, measuring instruments, radar, communications transmission, and switch equipment, and manufacturing equipment for electronic products. Requires products to have a label called the environmental protection use period (EPUP) that indicates the number of years it is expected to contain hazardous materials without causing environmental contamination (Chemsafetypro 2016).
  - *Health Canada*: Focuses on chemicals of concern including lead, mercury, BPA, and phthalates. Canada was the first government in the world to restrict BPA in products. They maintain a list of "Chemical Substances of Interest", which may lead to the restriction or banning of other materials (Health Canada 2016).
  - *California Green Chemistry (Safer Consumer Products)*: The California Green Chemistry Initiative was signed into law September 2008 and establishes a framework for regulating toxic substances based upon "life cycle thinking and green chemistry principles" (CalEPA 2010). Formal regulations took effect October 1, 2013, with the goals of reducing toxic chemicals in consumer products, creating new business opportunities in the emerging safer consumer products industry, and helping consumers and businesses identify what is in the products they buy for their families and

customers. The new regulations will require manufacturers to
seek safer alternatives to harmful chemical ingredients in widely
used products (DTSC 2016).

- *Washington Children's Safe Product Act*: Passed in 2008, this requires
  manufacturers of children's products sold in Washington to
  report if their product contains a "chemical of high concern to
  children." The CSPA also limits the amount of lead, cadmium,
  and phthalates allowed in children's products. In response
  to this regulation, several states have enacted similar laws.
  Vermont passed the Chemical Disclosure Program for Children's
  Products, Maine passed Safer Chemicals in Children's Products,
  and Oregon passed the Toxic-Free Kids Act (SWDE 2016).

- *California Proposition 65*: Proposition 65, the Safe Drinking Water and
  Toxic Enforcement Act of 1986, was enacted as a ballot initiative in
  November 1986. The aim of the Proposition is to protect California
  citizens and the state's drinking water sources from chemicals
  known to cause cancer, birth defects, or other reproductive harm;
  and to inform citizens about exposures to such chemicals. The rule
  requires notification and labels so that "no person in the course of
  doing business shall knowingly and intentionally expose any individual to a chemical known to the state (California) to cause cancer
  or reproductive toxicity without first giving a clear and reasonable
  warning" (CalEPA 2016).

- *Interstate Mercury Education & Reduction Clearinghouse (IMERC)*:
  In 2001, the United States Northeast Waste Management Officials'
  Association (NEWMOA) launched the Interstate Mercury Education
  and Reduction Clearinghouse (IMERC). Anyone who offers to sell,
  sells, or distributes a mercury-added product in a state covered by
  this rule is required to complete a notification form and submit it
  to IMERC. This information informs the public on which products
  have mercury so that its use may be minimized (NEWMOA 2016).

- *Globally Harmonized System of Classification and Labeling of Chemicals
  (GHS)*: "The GHS is a system for standardizing and harmonizing
  the classification and labeling of chemicals." It requires governments around the world to conform their hazard classification
  and communication rules to a common system, regardless of the
  country. This initiative will require changes in the way hazards
  of chemicals are defined and communicated. Consistency will be
  achieved globally in determining health, physical, and environmental hazards as well as communicating hazards through icons
  and chemical Safety Data Sheets. For example, countries have different definitions and icons for "flammable." With GHS, the icons
  and definitions would be the same regardless of the country you
  operate in (OSHA 2016).

There are many more regulations aside from these that have some form of labeling, restriction, or reporting on the toxic materials used in products. The intent here is not to have an exhaustive list, but to generally inform on the growing amount of regulations that manufacturers must be cognizant of to insure a compliant product on the global market.

Examples of Global Product Stewardship Regulations

- **Packaging regulations**—EU Packaging Directive and various others require size reduction, reduction of toxic metals, take-back
- **Chemical use restrictions & notifications**—REACH, RoHS, chemical specific bans e.g. Health Canada BPA ban, IMERC mercury use notification, CA Proposition 65 notifications, GHS
- **Extended Producer Responsibility**—WEEE, Battery, Brazil EPR, etc. require producers to take back products at their end of life

---

Extended Producer Responsibility

More governments are mandating that manufacturers take responsibility for their products at the end of their useful life and a category of regulations called extended producer responsibility have been developed. One of the first and most comprehensive regulations of this type is the **European Union Directive** 2012/19/EU known as *waste electrical and electronic equipment* **(WEEE)**. First enacted in 2003 (2002/96/EC), WEEE requires manufacturers to take responsibility for the recycling of equipment at its end of life. The equipment covered by the regulation includes large and small household appliances; IT and telecommunications equipment; consumer, toys, leisure, and sports equipment; lighting equipment; electrical and electronics tools, monitoring and control instruments, and automatic dispensers; and medical devices.

Producers of electrical and electronic equipment are required to set up and pay for collection points where WEEE can be brought to be recycled, and are responsible for the costs of collection, treatment, recovery, and disposal. Equipment is required to be designed to be recycled and must include guidelines for how to recycle the equipment and be labeled to indicate that it is WEEE (Europa 2016).

The US state of **California** also has a **WEEE** regulation: SB 20. It only applies to CRT, LCD, and plasma screens larger than four inches. A fee to cover recycling is paid for by consumers at the point of purchase. Manufacturers must notify retailers and the California Integrated Waste Management Board (CIWMB) when a device is subject to the recycling fee and provide consumer information on how to recycle the products. Annual reports must be filed with the Board indicating the total amount of hazardous substances in each device, the efforts to reduce hazardous materials, use of recyclable materials, and efforts to design more environmentally friendly products (CalRecycle 2015). California is not the only US state that has electronic equipment take-back requirements. In 2016, there were 25 US states that had some sort of electronic take-back requirements (Electronics TakeBack Coalition 2016).

Another component of electronic products that has caused environmental harm is batteries. Batteries contain toxic metals like lead, mercury, and cadmium and if improperly disposed of, cause environmental pollution. The **European Union Battery Directive** 2006/66/EC requires the collection and recycling of waste batteries and accumulators at their end of life. In addition, product design must enable batteries to be easily removed to facilitate recycling and include instructions on removing the batteries. Batteries must be labeled with a symbol indicating that they are to be recycled. Product manufacturers must cover the cost of collecting and recycling industrial, automotive, and portable batteries and accumulators, as well as the costs to inform the public of the recycling schemes (EUR-Lex 2014b).

On March 30, 2010, **Brazil Resolution SMA-024** was enacted. This rule is in essence a producer responsibility regulation. It requires some manufacturers, importers, and distributors to take responsibility for the post-consumption waste that their products generate. Products covered by the Resolution include: automobile oil filters; automobile oil containers; fluorescent bulbs; automobile batteries; tires; electro-electronic products; and primary, secondary, and tertiary packaging of foods and beverages, personal hygiene products, cleaning products, and durable consumer goods.

Manufacturers, importers, and distributors of these products must in partnerships or separately: "maintain collection posts for used products; to inform consumers of the need to return end-of-life products; to meet collection quotas; to report on the quantities collected; and to ensure that collected products are recycled, burned for energy, or otherwise disposed in a manner approved by CETESB" (the Brazilian environmental agency) (Beveridge & Diamond 2011).

Numerous EPR regulations have been developed in the United States. As seen in Figure 3.2, requirements have been established in many states for a variety of products from electronics to carpet, paint, and pharmaceuticals. These regulations are cropping up throughout the United States, but this is truly a global phenomenon that is building momentum.

The development of regulations for take-back of products and forcing companies to be responsible for the end of life of their products is becoming more popular with legislators. This policy initiative enables products to be properly managed when they have lost their usefulness and typically at a cost that is borne by manufacturers, which adds to its attractiveness. We should

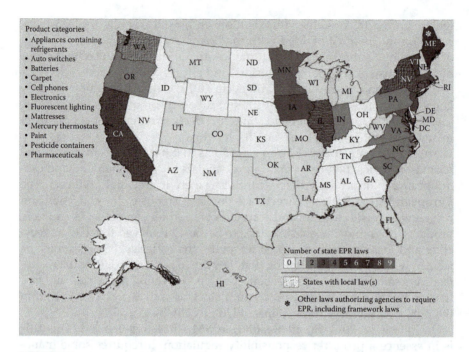

Product categories
• Appliances containing
  refrigerants
• Auto switches
• Batteries
• Carpet
• Cell phones
• Electronics
• Fluorescent lighting
• Mattresses
• Mercury thermostats
• Paint
• Pesticide containers
• Pharmaceuticals

Number of state EPR laws
0 1 2 3 4 5 6 7 8 9
States with local law(s)
* Other laws authorizing agencies to require
  EPR, including framework laws

FIGURE 3.2
EPR requirements throughout the United States. (Courtesy of Product Stewardship Institute, Inc., Boston, MA.)

expect to see more of this type of regulation in all markets throughout the world; this is especially true considering the new focus on developing a circular economy.

Supply-Chain Issues

Focus on environmental and social issues deep in the supply chain is a new regulatory initiative. Manufacturers are being held accountable not just for their own operations, but also that of their suppliers. It is becoming increasingly important to know what's in your product. This goes beyond the materials used in the product and includes social issues like the working conditions at suppliers, how suppliers' employees are treated, and political conditions of where raw materials are sourced.

In California SB 657, the California Transparency in Supply Chains Act requires companies doing business in California with more than $100 million in gross receipts to report on efforts to eradicate slavery and human trafficking from its supply chain. The reporting requirements are very broad and go into areas that governments never before considered. Disclosures of activities to insure that human trafficking is not occurring in a product's supply chain must be reported on the company's website. The information requires disclosure if the business:

1. Engages in verification of supply chains to address human trafficking and slavery and uses third parties in that process
2. Conducts independent, unannounced audits of suppliers to ensure compliance with company standards on trafficking and slavery
3. Requires direct suppliers to certify that materials incorporated in their products comply with the laws regarding slavery and human trafficking of the country or countries in which they do business
4. Maintains internal accountability standards for employees and contractors failing to meet company standards on slavery and trafficking
5. Provides both managerial and non-managerial employees with training on mitigating risks of slavery and trafficking in supply chains (Foley & Lardner 2011)

This view into a supply chain is adding more scrutiny to every step of bringing a product to market. Governments are responding to concerns with the conditions of the suppliers used for making products, and are holding manufacturers accountable to insure that their product does not enable horrendous conditions.

Another similar initiative is the use of conflict minerals. If someone would have told me that I would be working on **conflict minerals** as part of our product stewardship program, I would have thought that had nothing to do with our program. Now it is a frequently discussed product stewardship concern. Various products contain minerals that come from areas where there is conflict, such as Africa. Cell phones, laptop computers, televisions, medical equipment, and many other products contain these minerals.

The US government through the "conflict minerals" provision of the Dodd-Frank Act (Section 1502) will require companies to publicly disclose if they use these materials. Section 1502(e) (4) of the Dodd-Frank Act defines "conflict mineral" as cassiterite (tin), wolframite (tungsten), coltan (tantalum), and gold (gold) or their derivatives. Public companies are to report on minerals used in their manufactured goods that originated in war-torn Congo or adjoining countries in Africa. "The law is intended to address widespread corruption, human abuse and genocide in that region by imposing supply-chain due diligence on manufacturers that use in their products 'conflict minerals' that fund groups responsible for the atrocities" (Corporate Law Report 2011).

Another area that affects the sourcing of products in a significant way is bio-piracy, also known as access and benefit sharing (ABS) regulations. The Nagoya Protocol on Access to Genetic Resources and the Fair and Equitable Sharing of Benefits Arising from their Utilization was an outcome of the United Nations Convention on Biological Diversity. The idea of this regulation is to consider "the need to share costs and benefits between developed and developing countries." In the past, developing countries have had genetic resource like the use of unique plant-based raw materials used for consumer products, like natural ingredients and even pharmaceutical ingredients, removed from their country without permission or even compensation. Approvals and benefits are to occur for the use of these raw materials. Benefits may be monetary or non-monetary such as royalties and the sharing of research results. The Nagoya Protocol creates greater legal certainty and transparency for both providers and users of genetic resources by sharing the benefit of these resources and making it more predictable for the users of these materials. If you are a product manufacturer and are using ingredients such as natural or botanical extracts which are unique to a specific country, you may be subject to this new regulation (CBD 2016).

This new focus on the supply chain is a continuation of a trend that requires product manufactures to know the minutest details about how products are brought to market. Doing this can be extremely difficult since parts that go into products come from several tiers of suppliers, going back to the mine where minerals came from is not readily available information. Consider the many materials that go into one part, and that some products have thousands of parts. Now try to understand where all of the materials came from. This is a daunting task; nevertheless, supply-chain regulations are most likely going to increase and cover areas that today we cannot even imagine.

Company Management Systems for Product Regulations

With all of the new regulatory requirements that apply to products, how does a global company insure compliance? As mentioned earlier, more and more businesses are developing and marketing products that are sold globally. Having a robust system to insure that products being developed are meeting national standards is becoming increasingly more important. Management systems must be put in place for product development groups both in-house and for third-party partners, to insure compliance. As noted above, there are an enormous set of requirements that must be complied with.

One approach to manage this complexity is the use of company developed standards. A good example of this is Hewlett-Packard's (HP) General Specification for the Environment. To insure that products meet the global rules, it requires compliance to this standard in all contracts for design, manufacture, or purchase of HP brand products. This includes subassemblies, parts, materials, components, batteries, and packaging that are incorporated into HP brand products.

The standard has 102 pages of requirements. There is a section that addresses "Substances and materials requirements." HP restricts or prohibits certain materials in their parts and products. Some examples include: Halogenated flame retardants and polyvinyl chloride (PVC) limits of 0.1% (1000 ppm), hexavalent chromium and its compounds in metallic applications must not be present, mercury and its compounds 0.1% (1000 ppm), PVC in external case plastic parts must not be present. Mercury must not be intentionally added in any battery, and it must not contain more than 0.0001% (1 ppm) mercury by weight. Tributyl tin, triphenyl tin, tributyl tin oxide must not be used in parts, components, materials, or products.

The use of wood is also addressed. Parts, components, materials, and products must not contain any wood material or wild plant material that was illegally sourced from its country of origin. Heavy metals must be managed in packaging; the sum concentration of incidental lead, mercury, cadmium, and hexavalent chromium may not be greater than 0.01% (100 ppm) by weight. Product stewardship regulations are also addressed. Compliance must be achieved with any applicable regulations and documentation must be maintained. Examples include, China RoHS User Documentation, Korea e-Standby User Instructions for Personal Computers, Product End of Life Declarations and many, many more requirements (HP 2015).

The use of a management system such as this HP Standard is a good example of the type of processes that have to be put in place to insure developers and parts suppliers are providing compliant products onto the global marketplace. The fact that this document is over 100 pages long speaks to the difficulties of addressing all the requirements out there. One thing is for sure: Companies must develop some mechanism to track and assist R&D groups and suppliers on how to comply with the multitude of global requirements.

Managing Risk

As we have seen, regulations can have a huge impact on a company's ability to compete in the marketplace. Corporations are always trying to minimize their business risk and have adopted various methods to mitigate and anticipate adverse situations that can interrupt or even stop products from being sold. One of the most difficult situations to be in is when an issue comes out of nowhere, and you're completely blindsided by it. Even worst is having to defend yourselves in the court of public opinion because of it. As one of my company's business leaders used to say, "When you're explaining, you're losing."

As one of my companies business leaders used to say;
"When you're explaining—you're losing."

Business managers like to think that if you have good scientific evidence on your side, this is all you need to prove your point of view is correct. However, time after time we have seen that perception trumps scientific evidence. Once an issue gets out into the public in a widely distributed report, whether it's based on scientific evidence or not, it becomes very difficult to dispute after the fact. A good way of looking at risk associated with a product is to understand that risk = hazard + outrage (Blake 1995).

RISK = HAZARD + OUTRAGE

The leading factor that seems to bring public distrust of any industry is a lack of transparency. The public is leery of firms that don't appear forthright with information and are not telling them all they know about an issue.

A good example of an issue that has generated a lot of angst for some companies is genetically modified organisms (GMO). When the use of GMO crops first came to the public eye, there was outrage. Cries of Franken-food and environmental-group protests were prevalent. However, GMO crops have the possibility to address food shortages by being resistant to bad weather, insects or viruses (HGP 2008). Regardless of where you stand on the issue, it makes you wonder if the companies that developed these GMO crops had an emerging issues management program—did they see the outrage coming? Could they have been more transparent? Would it have been possible to have reached out to the opposition groups and share data? I don't have the answers to this, but a robust emerging-issues process would have given this issue a better chance of not being the firestorm that it was, and continues to be.

Factors that decrease risk	Factors that increase risk
Voluntary	Imposed
Control	Lack of control
Fair	Unfair
Ordinary	Memorable
Not dreaded	Dreaded
Natural	Technological, artificial
Certain	Uncertain
Familiar	Unfamiliar
Morally acceptable	Morally unacceptable
Trustworthy source	Untrustworthy source

Source: Blake 1995.

Corporate Reputation

During my career at Johnson & Johnson we had an exceptional CEO named Ralph Larsen. At one of our executive leadership meetings he reinforced the importance of our corporate reputation. Our name was not a trademark but a trust mark, and reputation is like a bank account, you make deposits and there are withdrawals, therefore we better make sure that we have a lot of deposits in our account to insure that we can withstand all assaults that will come. This sentiment aligns well with the principle's espoused in the book entitled, *The 18 Immutable Laws of Corporate Reputation.*

Similar to what Mr. Larsen said, it speaks to reputation being like a life preserver and is a "tailwind when you have opportunity." One of the laws is to "learn to play to many audiences," and one of those audiences is environmental groups. An example of doing this right was when fast food companies were pressured on environmental issues and nutrition. Instead of fostering an adversarial relationship, they worked with the pressure groups and developed practices that treated the farm animals that were used for food in their stores more ethically. This was a smart move and garnered positive comments from the groups that were putting pressure on them. I have personally been involved in working with various environmental groups and when you sit down and listen, you find out that there are common issues that usually can be worked through.

A good concept advocated in the above mentioned book, which is relevant to product-related concerns, is to get in front of the issues. When a controversial issue is raised against your company, it's usually a good idea to lead and proactively address the issue and speak publicly about what you're doing about it. If you do not take this tack, you may find your company on the wrong end of a boycott or taking a major reputational hit. Another "law" is to "control the Internet before it controls you." This concept connects to the proactive approach; we see many initial studies or campaigns initiated on the Internet. It's a wise move to monitor the blogs and websites of key environmental groups and influencers. Being vigilant and taking every perceived threat seriously is a good way to stay ahead of the curve and manage emerging issues that can affect a company's reputation (Alsop 2004).

Emerging Issues Management

It is possible to get signals on emerging issues that can result in business risk and appropriately influence their development. An issue takes a certain path before it becomes a regulation or a public relations nightmare. There are opportunities to address issues when they are in the "anticipatory" stage. Monitoring the warning signs coming from NGO reports, blogs, web postings, and research reports are imperative to getting a read on any developing business risk.

Emerging issues must be put through a filter that considers:

1. Potential business impact (both financial and reputational)
2. The time it will take to become a "crisis" and hit the public in the form of a news report or a regulation

Resources, such as the commissioning of teams of experts to study and recommend actions to management, should be deployed on the issues that have the greatest potential to (negatively) impact the business. Actions are most effective if they occur in the early stages of the "anticipatory" stage; see the chart in Figure 3.3.

Some of the outcomes of an effective emerging issues process could include:

- Commissioning research on topics to develop science.
- Papers developed and posted in peer-reviewed scientific journals to develop the science on the new issue.
- Presentations of data and perspectives at scientific conferences.
- White papers and position papers that discuss the issue and what the organizations' understanding and positions are on the issue.
- Communications in the form of meetings with NGOs or government agencies to share knowledge and points of view on issues.
- Development of company guidelines and standards.

An emerging issues process must have a strong connection to the company's government affairs group since they see what legislatures are considering first. Granted, this is much further down the path towards an issue becoming law, but it is critical to have sound scientific principles, front of mind, when regulations are being formed. Working with legislatures in the earliest stages possible is also necessary to prevent unnecessary regulatory burdens that do not add value.

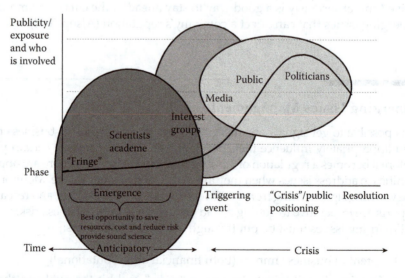

FIGURE 3.3
The general life cycle of an issue.

Examples of Emerging Issues

What type of issues are being monitored by companies today that can have significant business risk in the future? It depends a lot on the type of industry, but there are some issues that can affect quite a few different companies.

At Johnson & Johnson, we initiated an environmental health & safety (EHS) emerging issues process in 1998; it has been an extremely helpful process for evaluating the landscape, influencing potential regulation, and mitigating risk. In the formative days of the Emerging Issues Committee, we tagged such topics as the use of PVC, endocrine disrupting compounds, and the availability of freshwater. Over time, other topics that surfaced as significant issues included the presence of very small concentrations of pharmaceuticals and other consumer products in water, nanotechnology, bio-monitoring, and one-off chemicals being the target of pressure, such as bisphenol A, triclosan, parabens, phthalates, among others.

I can draw on my own experience to share a few examples of how an emerging issues process works. Let's take **PVC**—when it first arrived on our radar, there were no regulations prohibiting its use. However, environmental groups were petitioning companies to stop using PVC because they claimed it was the worst plastic to use because of manufacturing, and end-of-life concerns. Work on this issue resulted in the development of studies, a white paper position on PVC, and eventually, adoption of a public-facing sustainability goal to significantly reduce and in some cases, completely eliminate PVC from company packaging (to date over 3000 tons of PVC have been removed from Johnson & Johnson packaging). This approach has paid dividends because our customers are asking for products and packaging free of PVC and we are ahead on this issue because of our emerging issues process.

Nanotechnology is an example of an emerging issue. Governments are trying to determine the best way to address the use of compounds that are very common, but now can be milled down to nanometer size (nano size = 1 to 100 nanometers in dimension). Questions have been raised as to whether smaller sized materials will react differently in the environment or cause significant human health issues compared to larger-sized particles.

Some Current Emerging Issues

- **Biomonitoring**—finding of trace chemicals in human blood and body fluids
- **Chemicals of concern**—several individual chemicals being pressured by NGOs as harmful to human health & the environment, for example, BPA, triclosan, phthalates, DEHP, etc.

- **Chemical mixtures**—minute concentrations of chemicals in the environment and their collective impact
- **Climate Change**—impacts of climate change on raw material supplies and other potential business disruption
- **Endocrine disrupting chemicals (EDCs)**—fate of EDCs in the environment
- **Nanotechnology**—1 to 100 nano meters in dimension materials impact on human health and the environment
- **PPCPs**—trace amount of personal care products and pharmaceuticals found in the environment
- **Water Scarcity**—availability of safe and secure water supply

To address this concern, our emerging issues committee for nanotechnology partnered with other organizations to fund research on the environmental and human health impacts of these smaller particles. A safety guideline was developed to insure that new concerns are evaluated for any research or handling of nano-sized materials, and proper protection of employees are in place. Guidelines on the responsible use of nanotechnology were also developed that require risk assessments, consideration of social and ethical issues, marketing practices, and notification of suppliers. Further, our scientists met with government officials to share our knowledge and presented data at toxicology conferences.

Discussing Emerging Issues Publicly

As mentioned above, an important aspect to preventing outrage on emerging issues is dialog. It is increasingly common to see companies discuss dilemmas publicly. By doing so, it lets stakeholders know where they stand and that they are paying attention to issues they are concerned with. A good example of this is on Baxter's website.

Being a medical products manufacturer, Baxter had received some adverse publicity about the use of PVC in some of their products. The use of PVC is discussed on their website through a Materials Use Position Statement. This statement addresses exactly how the company sees the use of PVC in medical products and what they are doing to address customer concerns regarding this material.

Baxter Addresses PVC

"PVC has a long history of use in a variety of medical products, such as contact lenses, intravenous bags, oxygen tents and catheters. These products have undergone strict regulatory review by many government and independent health agencies throughout the world, including the U.S. Food and Drug Administration. The safety of these materials has been confirmed by more than 40 years of use with approximately eight billion patient days of acute and chronic exposure without report of significant adverse effects. Environmental questions relating to the incineration of PVC are being addressed with modern pollution control technologies that can reduce, for instance, dioxin emissions up to 99.9%.

While PVC is a widely used material that can consistently meet the rigorous requirements for medical products, it may not be appropriate for all clinical applications. To meet the preferences of some customers and address drug compatibility issues in specific clinical applications, Baxter offers a portfolio of intravenous medications, parenteral nutrition solutions, injectable drugs, biopharmaceuticals, IV sets and access devices and other products that use or are contained in non-PVC materials or non-DEHP [di-(2-ethylhexyl)phthalate], a common component of PVC, materials" (Baxter 2016).

This is a very good example of publicly stating a position on a topic of concern to stakeholders and demonstrating that consideration on the topic was made and non-PVC alternatives are being pursued.

Conclusion

Environmental regulatory requirements for products are increasing in all regions of the world. Governments are realizing that they can address many environmental and human health issues by focusing on products. Many have started in Europe and have inspired other governments to write similar regulatory requirements, such as REACH, RoHS, WEEE, and packaging regulations. Besides making regulatory compliance more complex, these expanding requirements are becoming a significant driver of greener product development.

Having to consider requirements that affect product design of packaging, toxic material use in products, and producer responsibility requirements for product end-of-life management all result in forcing greener product design. Some companies have enacted robust management systems to insure that product developers and suppliers are aware of and anticipating regulatory requirement in their design.

As these requirements become more important for products that are marketed globally, manufacturers are trying to track and influence the development of issues that can affect products marketability through emerging issues processes. It is evident that the increase of environmental product legislation will make tracking and complying with these new requirements a critical part of bringing greener products to market.

References

Alsop, Ronald J. 2004. *The 18 Immutable Laws of Corporate Reputation*. Free Press, New York, p. 17, 45, 75, 134.

Baxter. 2016. Position Statement—Material Use. http://www.baxter.com/press_room/positions_policies/pvc_position.html (Accessed December 29, 2016).

Beveridge & Diamond. 2011. Brazil Highlights. http://www.bdlaw.com/news-859.html (Accessed December 29, 2016).

Blake, Elinor R. 1995. Understanding Outrage: How Scientists Can Help Bridge the Risk Perception Gap. *Environmental Health Perspectives* 103, 123–125.

CalEPA. 2010. State Releases Proposed Green Chemistry Regulation. http://www.dtsc.ca.gov/PressRoom/upload/News-Release-T-09-10.pdf (Accessed December 29, 2016).

CalEPA. 2016. Proposition 65. http://oehha.ca.gov/prop65/p65faq.html (Accessed December 11, 2016).

California Department of Toxic Substances Control (DTSC). 2016. Safer Consumer Products. http://www.dtsc.ca.gov/SCP/index.cfm (Accessed December 9, 2016).

CalRecycle. 2015. Electronic Waste Recycling Act of 2003. 2015. April 9. http://www.calrecycle.ca.gov/electronics/act2003/ (Accessed January 3, 2017).

CBD. 2016. Convention on Biological Diversity. https://www.cbd.int/abs/ (Accessed December 17, 2016).

Chemsafetypro. 2016. China RoHS2 2016. http://www.chemsafetypro.com/Topics/Restriction/China_RoHS_2_vs_EU_RoHS_2.html (Accessed December 17, 2016).

CIRS. 2014. China New Substance Notification in China—China REACH. http://www.cirs-reach.com/China_Chemical_Regulation/IECSC_China_REACH_China_New_Chemical_Registration.html (Accessed December 30, 2016).

Corporate Law Report. 2011. Conflict Minerals and the Dodd-Frank Act: Long-Distance Accountability? February 11. http://corporatelaw.jdsupra.com/post/conflict-minerals-dodd-frank (Accessed December 30, 2016).

Department for Business Innovation and Skills (DBIS). 2014. *RoHS*. https://www.gov.uk/guidance/rohs-compliance-and-guidance (Accessed December 11, 2016).

Electronics Take Back Coalition. 2016. State Legislation. http://www.electronicstakeback.com/promote-good-laws/state-legislation/ (Accessed December 11, 2016).

EUR-Lex. 2014a. Disposal of Spent Batteries and Accumulators. May 19. http://eur-lex.europa.eu/legal-content/EN/TXT/?uri=URISERV:l21202 (Accessed December 30, 2016).

EUR-Lex. 2014b. Packaging and Packaging Waste. November 02. http://eur-lex.europa.eu/legal-content/EN/TXT/?uri=URISERV:l21207 (Accessed January 3, 2017).

Europa. 2016. Waste Electrical and Electronic Equipment. September 6. http://ec.europa.eu/environment/waste/weee/index_en.htm (Accessed December 30, 2016).

European Commission. 2016. *REACH*. http://ec.europa.eu/environment/chemicals/reach/reach_en.htm (Accessed December 10, 2016).

Foley & Lardner. 2011. New California Employment Regulations: Is the "Reprieve" Over? http://www.laboremploymentperspectives.com/new-california-employment-regulations-is-the-reprieve-over/ (Accessed January 3, 2017).

Health Canada. 2016. Chemical Substances. December 07. http://www.chemicalsubstanceschimiques.gc.ca/index-eng.php (Accessed December 11, 2016).

Hewlett-Packard Development Company, L.P. 2015. HP Standard 011 General Specification for the Environment. http://www.hp.com/hpinfo/global citizenship/environment/pdf/gse.pdf (Accessed December 30, 2016).

Human Genome Project (HGP). 2008. Genetically Modified Foods and Organisms. November 05. http://theliteratesims.net/eng1bM/Readings/gmfoodsand organisms.pdf (Accessed December 28, 2016).

NEWMOA. 2016. Interstate Mercury Education & Reduction Clearinghouse (IMERC). http://www.newmoa.org/prevention/mercury/imerc.cfm (Accessed December 11, 2016).

Occupational Safety and Health Administration (OSHA). 2016. Globally Harmonized System of Classification & Labeling. https://www.osha.gov/dsg/hazcom/global.html (Accessed December 11, 2016).

State of Oregon DEQ (ODEQ). 2005. International Packaging Regulations. http://www.deq.state.or.us/lq/pubs/docs/sw/packaging/intlpkgregulations.pdf (Accessed January 14, 2017).

State of Washington Department of Ecology (SWDE). 2016. Children's Safe Product Act. http://www.ecy.wa.gov/programs/hwtr/RTT/cspa/index.html (Accessed December 9, 2016).

Section II

Making Greener Products

4

Greener Product Design Examples

Ecomagination

Perhaps the most prominent and successful greener product initiative is GE's Ecomagination. An evaluation of GE's program can tell us a lot about the elements of a successful greener product development program. We also will get a perspective on how a firm with diversified products, from microwaves and dishwashers to medical imaging equipment to windmills and locomotives, addresses sustainable design.

Ecomagination is a well-rounded top-down initiative that has been given significant attention by GE management. You would be hard pressed not to have heard of this program since the company has used television commercials, print advertisements, and digital marketing to communicate their greener product offerings. The company tags products that have improved environmental performance as *Ecomagination*, distributes reports and brochures, and maintains a dedicated website, and their CEO is very public in speaking about the financial and environmental benefits of this initiative.

Development of Ecomagination

When it was first developed, the sustainability consulting firm GreenOrder was used to help develop the criteria by which to judge products, and a corporate team was established consisting of legal counsel, environmental health and safety, and marketing representatives to evaluate which products should go into the portfolio. GreenOrder would verify product information and the marketing claims to substantiate the environmental benefits of products. The characteristics considered to designate a product as Ecomagination include energy use, greenhouse gas emissions, water use, and the ability to offer financial benefits to their customers (Iannuzzi 2012).

To lend further credibility to GE's greener product approach, an independent advisory board was set up. The board consisted of members from non-governmental organizations (NGOs) like the World Resources Institute, Ceres, and academic institutions, Massachusetts Institute of Technology, and University College of London. Through my research of this program, I have not been able to find anything that describes the methodology used to determine how a product meets the Ecomagination standard. However, as we will see later in the description of some of the products, there are obvious environmental benefits (Iannuzzi 2012).

The greatest strength of Ecomagination is that it is represented as a business initiative, not an environmental one.

GE currently (2017) verifies product information and the marketing claims to substantiate the environmental benefits of products through external advisors such as the Executive Director of the Institute for Human Rights and Business, the Director of Corporate Social Responsibility of the Harvard Kennedy School, and the Senior Fellow of the Global Green Growth Institute and International Sustainability Development. A Sustainability Steering Committee that consists of leaders across the company with subject matter expertise also oversees the program (GE Annual Report 2015).

Perhaps the greatest strength of this program is that it is represented as a business initiative, not an environmental one. GE describes this program in business terms, profits from the products sold and customer desires are met. "Ecomagination is a business initiative to help meet customers' demand for more energy-efficient products and to drive reliable growth for GE." There is a focus on helping customers and society meet the environmental challenges of the day. By investing in cleaner technology and business innovation, GE is committed to driving economic growth and reducing their overall environmental impact. However, GE makes it clear that addressing these issues is not an altruistic endeavor; in meeting these challenges, business units must generate "profitable growth for the company." Positioning this program as a business imperative almost guarantees its success. Executives from various GE business units have taken notice and developed and marketed Ecomagination products (Ecomagination Strategy 2016).

It's interesting to see that Ecomagination is positioned as meeting the world's need for energy efficiency. According to GE, the world's energy infrastructure must be transformed because it is obsolete and uses environmentally dirty technology. Cleaner, more reliable, and efficient solutions to support the energy needs of future generations are required. That's where Ecomagination products come in.

Ecomagination Product Characteristics considered include:

- Energy use
- Greenhouse gas emissions
- Water use
- Ability to offer financial benefits to their customers

Ecomagination was launched in 2005 and has increased in size and scope since its beginning. GE has made significant funding investments to demonstrate their commitment; they invested $2.3 billion in clean technology for research and development in 2015. A total of $17 billion has been invested in research and development between 2005 and 2015. They also committed another $8 billion for R&D over the next 4 years. Success is measured in dollars; in 2015, Ecomagination technologies and solutions generated $36 billion in revenue. Not only are there goals for Ecomagination product sales, it also sets higher growth targets for these products at double the rate of the overall company growth. This is a huge testimony to the benefits of making greener products.

Setting growth targets that are double the rate of other products in the portfolio breaks the old paradigm that green products are barely profitable. In fact, Ecomagination products are sold in 100 countries, and GE sees a growing demand across the world. As of 2015, the company has reached 12% reduction of GHG emissions and 17% reduction of freshwater usage. GE is continuing to invest in this program, setting 2020 goals of $10 billion in additional R&D spend, 20% greenhouse gas reduction, and 20% freshwater reduction. It appears that GE is on a pace to meet, if not exceed, their 2020 goal for these targets.

Over a Decade of Ecomagination Results (2005–2015)

$17B R&D Spend	$232B Revenue Generated	12% Greenhouse Gas Reduction	17% Reduction in Water Use	40 GW Clean Energy Installed	1 B Gal/day Wastewater Treated	$98 Million Fuel Savings

(GE Ecomagination 2017)

The best way to understand the effectiveness of this program is to evaluate a few examples of the products having achieved the designation. In California's Mojave Desert, EcoROTR is a 1.7-megawatt test turbine funded by Ecomagination to analyze wind tunnel data. The goal of the project was to harvest more wind, which is exactly what the EcoROTR is currently doing. This turbine, which looks like an UFO that has toothpicks protruding from it, has an overall 3% increase in performance than a typical wind turbine. A 3% increase in efficiency is a big deal for windmills; this product can change renewable energy globally and create a more sustainable way to harvest energy (The Road to ecoROTR 2016).

The world is constantly in search of more efficient renewable energy sources. GE engineers have been working on the world's most powerful gas turbine, which is also known as 9HA. This machine can easily outdo a steam turbine by generating enough electricity to power about 600,000 homes; it converts natural gas into electricity at 61% efficiency. This new technology also transmits analytical information to GE's Predix platform through hundreds of sensors found on the turbine. The 9HA is so efficient that the company already has about $1 billion in orders before it has officially launched—that's what Ecomagination is all about (What This Turbine Does, September 21, 2016).

Software can also be an Ecomagination product; the Predix platform is the world's first industrial app that will monitor and control analytical data from various industrial categories. Data are extremely important when trying to make technology more efficient; the app provides real-time streaming of different machines such as the 9HA and locomotives produced by GE. This platform will provide engineers with an opportunity to create apps for "aviation, agriculture, health care, manufacturing and transportation" and to connect vendors and customers to receive timely analytical results, which results in more efficiency and thus lower environmental impact (Get Your Software Kicks on Predix 2016).

Substantial water savings resulted from the installation of GE's reverse osmosis system and cooling water treatment process at Omnova Solutions' Green Bay Wisconsin facility. The benefits were realized by sending wastewater for cooling tower use instead of being discharged to the sewer. This resulted in eliminating an astounding 3.6 million gallons of water use and saving $120,000 per year, as well as recognition as an Ecomagination product. This water treatment process received a "Return on the Environment Award" from GE's power and water division for its efficiency (OMNOVA 2016).

Ecomagination products seem to be penetrating all aspects of society. Starbucks, for example, was looking for more viable, energy-efficient lighting for their stores. In a partnership with GE, highly efficient LED lighting was developed. As of September 2010, over 7,000 stores have installed the LED lighting. Compared to typical store lighting, this saves approximately 8,100 metric tons of CO_2 emissions, the equivalent of 1,600 cars on US highways. Each LED light that is used by Starbucks reduces energy costs and CO_2 emissions by about half a barrel of oil (How Starbucks Saves Millions 2016).

Key Attributes of Ecomagination

- Represented as a business initiative
- Strong CEO support
- Supported with billions of R&D dollars
- Third-party verified
- Has review board of external company advisors

As we can see from reviewing some of the Ecomagination products, there are evident benefits produced under this banner. The initiative's effectiveness is exemplified in the breadth of the products that have achieved the designation. The financial support and targets set by the company show that it is firmly behind this program, not only to help the world solve its environmental problems, but also to bring profits to the bottom line.

Timberland's Green Index®

An impressive greener product development program is by Timberland. This is the first company that has shown leadership in this sector by putting a customer-facing nutrition-like label on their product packaging, which they call the Green Index®. This index reflects the environmental impact that the product has (Timberland Green Index 2016) (Figure 4.1).

Timberland's obvious interest in greening their products is evident in reading their Green Index® report. The company is extremely transparent on how they evaluate their products and they go into great detail in describing their program that develops and evaluates products in their Green Index report. Three categories are used to rate products: climate impact, chemicals used, and resource consumption. These factors are measured on a scale of 1–10.

FIGURE 4.1
Timberland's nutrition label via the Green Index® (Iannuzzi 2012).

Each factor gets a score and is divided by 3 to determine the environmental impact. Much can be learned by evaluating this innovative program.

Green Index® Rating Categories

Climate Impact
Greenhouse gases and emissions produced in making raw materials, and during footwear production, contribute to climate change.
Chemicals Used
Chemicals are used in material and footwear production to improve the performance and aesthetics of products.
Resource Consumption
Resource consumption is an intensive process that leaves a significant environmental footprint.

Using life-cycle assessment (LCA), Timberland evaluated their iconic yellow boot. This assessment was used to identify the areas that have the most impact from raw-material extraction, production, and transportation. The results indicated that the largest environmental impacts come from raw material extraction and production; this leads developers to focus on reducing impact of manufacturing and raw materials. By focusing their improvement efforts on the hot spots from the LCA, improvements can be targeted on the most important life-cycle impacts. When developing the label, they had to make some decisions on what not to measure and report on. For instance, transportation of footwear is not a focus area because it is an insignificant impact, less than 5% of the total climate impact.

Let's look deeper into the three metrics for the Green Index® to get a clear understanding of how they are used in product evaluation. The climate metric includes the electricity data from manufacturing and raw material extraction which are converted into greenhouse gas emissions. The chemical metric focuses on the use of solvent-based adhesives and PVC in the shoe materials. The resource metric considers the weight of recycled, renewable, and organic materials used in a shoe. Each category includes a calculation that results in a 0 to 10 score. These three aspects are monitored, measured, and then given scores that evaluate the environmental impact of each product made. The 0 to 10 scale is used with 0 being the lowest environmental impact and 10 being the highest. The score is then averaged for a total score; each category summary is listed on the label, which is put on the product package, similar to a nutrition label (Timberland 2016).

Metrics like these are helpful for companies to drive improved performance and innovation. Timberland's developers compare scores of new footwear to the existing model to determine if they are making it greener. It should be noted that not all merchandise offered by Timberland has a Green Index® label.

A good example of how product improvement is communicated is the innovations that have been made to the Classic 6" Premium Waterproof Boot.

This boot uses less raw material and more recycled materials, which lowers their resources score. The boot has Primaloft® Eco Insulation, which uses synthetic fibers that are at least 50% recycled polyethylene terephthalate (PET). PET is the plastic used in making soda bottles, which Timberland recycles to use in the making of their boots. Timberland has also stated goals to use leather that comes from silver- or gold-rated tanneries. This shoe derives at least 50% of its leather from a facility that is rated Silver or higher. The Silver rating classifies the tannery based on energy use, waste production, and water treatment (Timberland 2016). Consumers interested in a greener product would be more comfortable purchasing this boot knowing that it was made with conscious environmental improvements.

Key Attributes of Timberland's Product Stewardship Program

- Use the label "Green Index" to communicate product sustainability
- Have Earthkeepers® line of greener products
- Use Timberland Environmental Product Standards to drive company-wide product improvements

---

In addition to the Earthkeepers® line of greener footwear, Timberland has created environmental standards called Timberland Environmental Product Standards (TEPS) for all Timberland product categories. By doing this, they have created meaningful environmental standards to "green up" their entire line of products. Through TEPS, Timberland has set goals for all footwear to contain at least one recycled material. This paves a way for their shoes to become even more eco-friendly in the future. Timberland is also pledging that all shoes and apparel will be PVC-free—one of the most undesirable plastics. The company also seeks to use leather that is sourced from tanneries that have a Gold or Silver rating from the Leather Working Group for following the best environmental practices. Finally, all Timberland's apparel will use cotton from the Better Cotton Initiative and will be 100% organic by year 2020 (New Product Standards 2016).

Using a combination of the Green Index with overarching product stewardship goals is a good method for driving product improvements. The Green Index helps product developers focus environmental improvements on the most important life-cycle impacts. The higher level product stewardship goals help emphasize and quantify corporate improvements to their products.

SC Johnson Greenlist™

SC Johnson is a privately held company based in Racine, Wisconsin, and is a global provider of consumer products. With over $8 billion in sales, their product lines include a variety of merchandise used in homes, such as Glade air fresheners, Ziploc plastic sandwich bags, household cleaners Pledge and Windex, and even the Raid pesticide brand. Manufactures of these product groups have been challenged by customers and NGOs to improve their formulas because of questionable ingredients that may have toxic effects on the environment and to the customers that use them. The market has been shifting with the introduction of newer products claiming to be more natural or green.

SC Johnson has been very public about focusing on greening their products. To positively affect product design, they made sustainability a company-wide initiative and not a single department's responsibility. The goal of creating the green list was to focus on making "better" options for products instead of focusing on taking out the "bad" ingredients. The primary way that products are greened is by continuously improving the raw materials used. The measurement system to drive the greener product improvements is a process called Greenlist™ (SC Johnson, 2015 Public Sustainability Report 2016).

SC Johnson developed Greenlist™ in 2001; this continuous improvement process rates each ingredient from 3 to 0. Ingredients are put into categories of Best (3), Better (2), Good (1), and the least desirable to use, 0-rated materials. Each ingredient is based on four to seven key criteria such as biodegradability and toxicity. The objective is to increase the score of a product ingredient's over time. The 0-rated materials are a specific area of focus and are only to be used if there is no other workable alternative.

It is very useful to have tools and objectives to enable product developers to develop greener products like Greenlist™; it makes it easier for positive decisions to be made. SC Johnson scientists are tasked with developing new products that use raw materials rated as Better or Best. As with most consumer product companies, there are times when product formulations are updated; in such a case, the scientists must include ingredients that have ratings equal to or higher than the original formula.

The Greenlist approach not only focuses on improving products but also focuses on supply-chain practices. The company puts a higher value on suppliers that demonstrate environmental responsibility through such programs as ISO14001 certification. Also, they have a focus on preventing deforestation through the sustainable sourcing of pulp, paper, packaging, and palm oil. Suppliers that help with this commitment will no doubt gain opportunity for more business (SC Johnson 2016).

Results of Greenlist™

The benefit of having a metric-based system to rate your products is that progress can clearly be demonstrated to stakeholders. Using 2000/2001 as a baseline in the reporting year 2015/2016, SC Johnson's product formulas improved the amount of what they consider the best ingredients (3 and 4 rated ingredients). Back in 2001, the company started at 18% "Better/Best" ingredients and is currently at 52%, indicating to all concerned that significant progress has been made. The SC Johnson Greenlist™ process score uses a four-step process: classifying raw material type; looking at supplier-provided data to identify any 0-rated, or restricted-use, materials; generating criteria scores; and calculating the overall material score.

Greenlist™ Four-Step Scoring Process

1. Raw material
2. Supplier-approved data
3. Generating criteria scores
4. Overall material score

Greenlist™ has also been responsible for encouraging product developers to remove chlorine-containing packaging from their products, including PVC plastic and bleached paperboard (Figure 4.2).

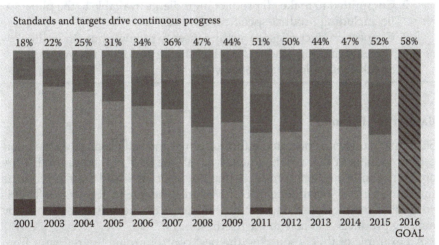

This chart shows SC Johnson's progress increasing the use of Better/Best ingredients as rated by the SC Johnson Greenlist™ process. Due to ongoing advances in measurement, changes in our product portfolio, and increasing numbers of ingredients and materials being measured, data may not always reflect apples-to-apples comparisons year on year. Also, scores after 2011 reflect combined ingredient and package scores. Prior to 2011, scores were ingredients only. Totals are rounded.

The SC Johnson Greenlist™ process uses a four-point scale:
■ 3 - Best ■ 2 - Better ■ 1 - Good ■ 0 - 0-rated materials

FIGURE 4.2
Greenlist™ progress for increasing the use of better and best ingredients.

One of the ways to generate more public trust in the environmental and health claims a company makes is to become more transparent about a product's ingredients. There have been a lot of NGO reports on the hazardous nature of household products, such as the type manufactured by SC Johnson. Taking a bold step, the company has committed to make all of its ingredients available to the public. "CEO Fisk Johnson explained... 'Making information about the ingredients in our products readily accessible and easy to understand helps our consumers know they can continue to trust our products'" (SC Johnson 2016).

SC Johnson, going the extra mile, has committed to list all ingredients, including fragrances and preservatives, on a dedicated website called whatsinsidescjohnson.com. This is a leadership position for the industry and a trend that I believe will eventually be a requirement. As of 2015, SC Johnson began disclosing the fragrances used in products and is committed to complete and total transparency of their products' ingredients (SC Johnson 2016).

Key Attributes of SC Johnson's Program

1. Greenlist™ metric-based ingredient improvement system
2. Committed to make all product ingredients available to the public, including product-specific fragrances
3. Connect product sales to worthy causes to promote human health
4. Focus on making products available to the bottom of the pyramid
5. Speak about dilemmas publicly

Base of the Pyramid

SC Johnson makes a concerted effort to source and sell products to the poorest individuals in the world, which they call the base of the pyramid. They are trying to bring forth products that are affordable and source products that can bring a social benefit to the poorest people in the world. This is admirable and a great example of how business can be a force for good—by making products available to the neediest.

An example of this is sourcing the pest control product pyrethrum from local suppliers in east Africa. Pyrethrum can be manufactured through a synthetic process; however, it can be derived naturally by extraction from chrysanthemums. To enable economic growth and a sustainably sourced raw material, SC Johnson is working with local suppliers in East Africa through a partnership with the U.S. Agency for International Development to improve farming techniques which could enable a more consistent supply. Through this effort, farmers in Africa are able to increase their crop yields and make more profit while providing a naturally derived ingredient.

New supply chains are being developed to bring products to those that traditionally could not afford them. This will enable pest control products into markets where it could help reduce insect-borne diseases like dengue fever and malaria. Another initiative that focuses on helping the base of the pyramid is SC Johnson and Cornell University's Center for Sustainable Global Enterprise's WOW™ club pilot. This three-year project started in 2012 with the goal to reduce transmission of malaria by creating a new business model that would bring bug repellents and insecticides to rural families and focus on helping low-income homemakers care for their homes and families (SC Johnson 2016).

Speaking about Dilemmas

NGOs have focused on low-level concentrations of potentially problematic ingredients in consumer products. Publicly discussing a point of view and sharing data can build trust in your product stewardship program. One of the ingredients of concern which is used in some of SC Johnson's products is phthalates. The one that has received the most negative focus is *diethyl phthalate* (DEP). This issue is discussed on their website, where they state that the evidence of numerous studies has indicated that DEP is safe for use, as is the whole class of phthalates. But since the company is committed to going above and beyond, they have decided to phase out the use of DEP in all their fragrances.

Anyone who has worked in the area of product sustainability will know that perception is reality in the eyes of the public. Conceding this fact, SC Johnson explains that although the scientific evidence indicates that there is no concern, they understand the consumer's desire to have products without materials of concern. So suppliers were asked to phase out DEP in all their fragrances.

Speaking out about *dilemmas* such as this builds public trust and lends credibility to discussions with NGOs. Sometimes difficult decisions have to be made based on customers' perceptions, even if company scientists disagree. Only leaders in product stewardship are willing to discuss sensitive issues such as product ingredients, publicly.

SC Johnson deploys a metric-based system that makes it easy for developers to focus product improvements and to report public progress. They have also connected their products to special causes that will benefit the poorest consumers. Besides greening up the existing line of products, a greener product line was brought to market based on natural ingredients. SC Johnson has taken a leadership position in voluntarily listing all product ingredients on their website.

Clorox Green Works®

Clorox, best known as the makers of bleach, also markets such products as plastic wrap, household cleaners, water filters, and charcoal for grilling food.

In recent years, they have become well known for developing eco-innovative products. The most meaningful way to improve a product's environmental performance is to focus on its greatest life-cycle impacts. When looking at consumer goods, manufacturers find that the most significant impacts are in the product itself not in the manufacturing plant. This is especially true when you place over one billion products in the marketplace every year.

An excellent example of developing and marketing a greener product came as a surprise to many. The entry of the Green Works® line was revolutionary and a real gamechanger for the household cleaning industry. Clorox, which is known for creating bleach and various other successful biocides, re-branded their image as safe and free of harsh chemicals. Green Works® products are based on a natural ingredients platform. When visiting a supermarket in the United States, you will see a prominent display of these natural, green cleaners.

Perhaps the greatest accomplishment of this product line is not that they have had so much success, but more in the way it vaulted greener products into the mainstream marketplace. The old stereotype of green products being ineffective and not of high quality was broken. Having a leading manufacturer of cleaning products introduce a natural product line enabled the public to embrace and feel safe purchasing greener products.

Clorox explains that the reason Green Works® products were developed was to meet customers demand for naturally derived cleaning products. Their evaluation of the market was right on and apparently their timing was perfect. In my opinion, the success of the Green Works® line paved the way for other brands, even those that were not previously sold in the mainstream grocery store, such as Seventh Generation.

Green Works Product Standards

- Made with plant and mineral-based cleaning ingredients
- Come from biodegradable ingredients that are naturally derived
- Are not tested on animals
- Use sustainable packaging whenever possible
- Are acknowledged by the EPA's Safer Choice program

Certification and Partnership

When Green Works was first developed, Clorox took two significant steps to bolster the confidence and trust of customers and stakeholders—receiving the U.S. Environmental Protection Agency (EPA) Design for the Environment (now called Safer Choice) product certification and getting the endorsement from a major environmental NGO, the Sierra Club. By doing this, their new green brand's "natural" image was reinforced (Iannuzzi 2012) (The Clorox Company, 2016 Integrated Annual Report 2016).

A smart move for Green Works® was partnering with the Sierra Club when it was introduced in 2008. This backing is significant because the Sierra Club is one of the best-known environmental groups in the world. Products displayed the Sierra Club logo which indicates its endorsement of the brand. The Sierra Club does not endorse Green Works anymore, but using their status to endorse their new line of green products *at the time* was a very smart, strategic marketing technique that gave the brand credibility.

The Safer Choice certification requires a stringent evaluation of the product's ingredients. Products that receive this certification have reached a level that indicates their ingredients have the lowest impacts in their product class. It is designed to protect the health of families and the environment. Though not widely recognized by consumers, the use of the Safer Choice logo on a product does go a long way toward defusing criticism by NGO watchdog groups and makes retailers more comfortable with touting the product as greener. Since the EPA uses a robust process of ingredient reviews and the use of predictive models, their certification quells claims of greenwashing.

An additional natural line of products Clorox acquired in 2007 is the iconic Burt's Bees®. Sold in mainstream supermarkets, it is a leading natural personal care brand in the United States that is also marketed internationally. Clorox has expanded Burt's business model, one that has a strong social responsibility component, where they give 10% of all sales to worthy causes. An example is their moisturizing lipstick with packaging that is made of 60% post-consumer recycled materials and it does not come in a box in order to eliminate excessive packaging (The Clorox Company 2016).

Clorox 2020 Eco Goals

1. Make product portfolio more cost-efficient and sustainable through improvements to 50% of product portfolio (vs. 2011).
2. Improve operations impact by reducing footprint intensity by 20% (vs. 2011).
3. 20% of sales growth driven by eco initiatives.
4. Environmental stewardship reputation is at CPG exemplar level.

Green Works® and Burt's Bees® have been a big success, but **what about Clorox's legacy products?** To address their main products, a corporate goal has been set to achieve sustainability improvements for 50% of their product portfolio by 2020 using 2011 as the baseline year. Clorox base products are engaging in sustainability improvements by meeting two or more of the following four criteria: (1) 5% or more reduction in product or packaging materials on a per-consumer-use basis, (2) an environmentally beneficial change to 10% or more of packaging of active ingredients on a per-consumer-use basis, (3) a 10% reduction in usage of water or energy by the consumer, and (4) an environmentally beneficial sourcing change to 20% or more of active ingredients or packaging on a per-consumer-use basis. In order to achieve this goal, improvements will have to be made to an extensive amount of products. Clorox has made sustainability improvements to 31% of their products between 2011 and 2016 (The Clorox Company 2016).

Examples of greening existing product lines include the development of the Glad® compostable trash bag made with renewable resources. Glad® has removed 6.5% of plastic used in their base trash bags—an equivalent of about 140 million fewer trash bags annually. These bags will be completely biodegradable and 100% compostable; they will disintegrate into compost fairly quickly to support the life of plants.

Having products that are inherently greener is another way to demonstrate a company's commitment to sustainability, enabling customers to reduce *their* environmental impacts. The Brita® water pitcher filter provides clean drinking water. Customers using this instead of disposable bottles can replace as many as 300 standard 16.9 oz. bottles of water. This is good not only for reducing plastic bottles but also for saving consumer's money. Clorox estimates that a Brita® pitcher filters 240 gallons of water per year for about 19 cents per day. This is a substantially lower cost than purchasing a bottle of water. In addition, consumers are able to recycle their Brita® water pitcher filters with Preserve®, who uses the recycled filters in a line of eco-friendly 100% recyclable products.

Key Attributes of Clorox Green Products Initiatives

- Development of the natural products line Green Works®
- Acquisition of Burt's Bees® green product line
- EPA's Safer Choice to bolster green credentials
- Set sustainable product and packaging goals for legacy products

---

A big impact area for consumer goods is packaging waste. Many Clorox company brands have moved to more sustainable packaging. Package redesign, material reduction, and increases in recycled content have led to measurable improvements. Clorox has made significant progress in this area, with 90% of product packages made of 100% recycled material and 100% post-consumer waste content used in most of their US retail display materials. Clorox® regular bleach has reduced the plastic in its bottles by more than 5 million pounds annually. Packaging objectives have been set for their entire portfolio; goals include making sure that 90% of all products use recyclable packaging, include recycling instructions on retail packaging, use only recycled or certified virgin fiber in packaging, and eliminating the use of PVC in all packaging (The Clorox Company 2016).

Another leadership position taken by Clorox was being the first in the cleaning industry to commit to voluntary disclosure of product ingredients. Work on this objective started in 2008 when all ingredients were listed on product labels for the Green Works® line. Traditional brand household and commercial cleaners had their ingredients listed on the corporate website in 2009. Clorox states that their priority is keeping families, pets, and the environment safe.

Clorox has been a trailblazer in bringing greener products to the mainstream consumer with the Green Works® line. Not only have greener product lines been brought to market but also legacy product lines are being greened up by setting corporate goals to improve the environmental profile of products and packages. The credentials of the products are bolstered with third-party certifications and donations of a portion of sales to good causes.

Johnson & Johnson

Johnson & Johnson (J&J) is the world's largest health-care product provider; it is also the company I know the most about since I have worked there for over 30 years. J&J has a very broad line of products, with sales in excess of $74 billion (2016). There are three main divisions: consumer products, medical devices, and pharmaceuticals. Some of the brands that may be familiar to you include Johnsons' baby products, Aveeno, Neutrogena, the Acuvue contact lens, over-the-counter pain medication Tylenol and Motrin, and in addition numerous medical devices like replacement hips and knees, sutures and various pharmaceutical products that treat migraines and rheumatoid arthritis to cancer. Developing product stewardship programs for a wide array of different products has its challenges.

I have been fortunate to work for a company that embraced the concept of product stewardship before it became a business imperative. In the 1990s, J&J's first product stewardship initiatives began with the public facing pollution prevention goals. These goals were primarily traditional footprint reduction goals at manufacturing facilities (e.g., water, waste, energy reductions) but there also was a packaging reduction goal, the first attempt at reducing the impacts of the product itself.

In the late 1990s, I was privileged to develop and lead the J&J design for the environment (DfE) program. This initiative covered all aspects of the product life cycle, from evaluation of raw materials to improvements of the manufacturing process, the product and package, emerging issues as well as end-of-life disposal issues. In 2000, the company started to develop and report on public facing sustainability goals every 5 years. The goals, called the Next-Generation Goals, included a requirement that all products and packages have an environmental impact analysis performed, considering ways in which impacts can be reduced.

Earthwards®

In 2009, the feeling was that the company needed to get more out of the DfE program and make it easier for development personnel to determine how to make a product greener and for marketing personnel to communicate the greener attributes of product improvements. This was the rationale behind the development of Earthwards®.

Earthwards® is a process that enables product-development teams to evaluate a product throughout its life cycle and identify areas where it can be improved, to lower its impact and increase social benefit. Tools and a scorecard have been developed to assist design teams to uncover improvements to reduce product impacts. Products that have been significantly improved can receive special recognition if they complete a life-cycle screen, meet prerequisites, and achieve the Earthwards® criteria.

The scorecard was developed through benchmarking leading companies through interviewing internal and external stakeholders, and by the guidance of a leading product stewardship consultant, Five Winds International. In addition, over the years we have solicited evaluation of the Earthwards® approach by experts from government, academia, business, and NGOs to make recommendations for improvements.

Process Steps

The Earthwards® approach consists of four steps.

1. *Meet product stewardship requirements:* New products must achieve regulatory compliance and deliver on Johnson & Johnson's high standards. Product teams are to answer a series of questions; examples include:
   - Are materials sourced from environmentally or culturally sensitive regions?
   - Identify 100% Watch List materials/ingredients in our products.
   - Have you done a review to assess where the product and packaging end up after use?

2. *Be reviewed for life-cycle impacts:* The life-cycle impacts of products are reviewed at the category level, and opportunities to drive improvements are considered at the design, procurement, manufacturing, and marketing stages of a product's development.

3. *Implement and validate improvements:* Product teams collaborate with sustainability experts to implement recommended improvements, and environmental marketing claims are reviewed and approved in accordance with applicable guidelines.

4. *Achieve Earthwards® recognition,* an honor celebrating our most innovative and improved products. If a product achieves at least three significant improvements across seven impact areas, a board of internal and external experts determines if the product warrants Earthwards® recognition (Johnson & Johnson 2017b) (Figure 4.3).

FIGURE 4.3
The Earthwards® approach (Earthwards® is a registered trademark of Johnson & Johnson. Reproduced with permission from Johnson & Johnson).

Scorecard Approval

If the product team believes that at least *three sustainable improvements* were made within seven categories, then they can submit the scorecard for verification to a review board consisting of internal J&J experts, legal counsel, public affairs, third parties including academics and an NGO. The board then makes a decision whether the product meets the Earthwards® criteria. If it does meet the criteria, the product receives the Earthwards® designation and is considered a recognized product. We are careful in the words we use to describe our most improved products and through much debate and thought came up with calling them Earthwards® **"recognized"** products. Since this is an internal company program, we felt the word "recognized" *would not* allow the Earthwards® logo or mark to be confused with a third-party eco-logo. If we used a word like "certified," that could confuse some customers into thinking it might be an eco-logo. As discussed in detail in Chapter 9, words matter when trying to avoid being accused of greenwashing.

The seven areas for improvement are:

1. Materials
   - Meet consumer needs with less material
   - Use more environmentally preferable material
2. Packaging
   - Reduce packaging
   - Use more sustainable packaging materials

3. Energy
- Create a less energy-intensive product
- Use more efficient manufacturing and distribution processes

4. Waste
- Reduce waste during manufacturing
- Recover more products for reuse or recycling

5. Water
- Generate a more water-efficient product
- Make manufacturing process more water-efficient

6. Innovation
- Initiate quantifiable environmental improvements in a product or process that has not been captured in another scorecard category

7. Social
- Use fair-trade materials, select socially responsible suppliers, or support causes with clear social/environmental benefits (Johnson & Johnson 2010) (Figure 4.4)

Earthwards™ Scorecard		
Pre-requisites		Achieved
	Know materials and ingredients in your product	
	Identify and plan to address J&J Watch List materials	
	Know where product and packaging end up after use	
	Know whether agricultural or mined ingredients come from culturally or environmentally sensitive regions	
Complete life cycle screen to **_identify priority goals_** for the product (see separate screening questions)		
Goals		
Materials	1 Meet consumer need with less material *or*	
	2 Use more environmentally preferred material (see list) *or*	
Packaging	3 Meet consumer need with less packaging *or*	
	4 Use more environmentally preferred packaging material (see list) *or*	
Energy	5 Make product more energy efficient in use *or*	
	6 Make manufacturing or distribution more energy efficient *or*	
Water	7 Make product more water efficient in use *or*	
	8 Make manufacturing more water efficient *or*	
Waste	9 Make product with less waste during manufacturing *or*	
	10 Recover more product, after use, for reuse or recycling	
Results	**Achieved all pre-requisites + three other goals?**	

FIGURE 4.4
Johnson & Johnson's Earthwards® scorecard.

Earthwards® Recognized Product Examples

Simponi® (Golimumab)

If you were suffering from rheumatoid arthritis, would you prefer your treatment injections to be once per month or three or four times a month? Simponi® is an innovative therapy that requires self-injections you can do in the comfort of your home. Not only is this therapy more convenient and less invasive for the patient but also it is much more sustainable than other treatments, using less materials and less energy.

Innovation

By requiring only 12 injections per year, patient needs are met using 36–61% less material. In the Unites States, a new sample distribution system includes a shipper that is returnable and reusable, and employs USDA certified bio-based cooling materials. The previous shipping container was single-use disposable.

Earthwards® has two objectives
1. Give clear line of site to R&D and marketing professionals on how to make a product greener.
2. Develop green marketing claims backed by science, facts, and data.

Packaging

The new shipper is 50% lighter and helps prevent disposal of more than 42,000 cubic feet of Styrofoam annually compared to the original shipping method (Iannuzzi 2012). This is a perfect example of an Earthwards® recognized product—one that is better for the customer (the patient in this case) and has improved environmental performance to a comparison product. I always say that Earthwards® has two objectives: (1) give a clear line of site to R&D and marketing professionals on how to make a product greener, and (2) develop green marketing claims backed by science, facts, and data.

Some other examples of Earthwards® recognized products include AVEENO® PURE RENEWAL™ Shampoo and Conditioner, which earned Earthwards® recognition as a result of reducing energy use by converting a transportation route from truck to rail and using the ingredient NATRASURF™ as part of its sulfate-free formula. NATRASURF™, the first personal care ingredient of its kind, is a potato starch-based polymeric surfactant with the same performance as traditional surfactants.

The Sterilmed® Trocar, a small instrument used in surgical procedures, is collected, remanufactured, and sold as a reprocessed device. The Sterilmed Trocar received Earthwards® recognition because it represents a 50% improvement in product waste versus an original single-use trocar device: Where an original trocar begins its life using raw materials and ends its life in a landfill or incinerator, the Sterilmed Trocar is collected, reprocessed, reused and then ends its life at a facility that creates energy out of waste.

ZYTIGA® is used to treat prostate cancer, which is the second most frequently diagnosed cancer in men and the fifth most common cancer overall. By applying green chemistry principles in the formulation of this product, Janssen (a J&J pharmaceutical company) doubled the process yield and realized significant reductions in raw material use, water use, and hazardous waste generation (Johnson & Johnson Strategic Framework 2017a).

Citizenship & Sustainability
2020 Product Stewardship Goals

J&J Established a 2020 Citizenship & Sustainability goal to fully integrate sustainable design solutions into product innovation processes. Some of the metrics are:

- **Goal:** 20% Johnson & Johnson revenue from Earthwards® Recognized Products
- **Goal:** Increase recyclability of consumer product packaging to 90+% in key markets through design for recyclability and partnerships

---

Earthwards® Results

Since its inception, Earthwards® has steadily grown, and at the end of 2016, there were 96 of the most improved "recognized products." To demonstrate impact of the initiative, sales of the recognized products are tracked. At the end of 2015, there was approximately $9.3 billion in revenue—roughly 13% of sales—from Earthwards®-recognized products.

Impacts are not only calculated by sales but on environmental improvements: at the end of 2015, approximately 3,600 metric tons (MTs) of reduction in **packaging**, 18,270 MTs of **materials** removed, 6,630 liters of less water used and 3,630 MTs of waste reduced. Improvements in products were made across the seven Earthwards® categories, covering *all* of the categories, as you can see from the list below.

Impact Area	Number of Improvements
Materials	85
Packaging	70
Energy	45
Waste	19
Water	18
Social	35
Innovation	25

Source: Johnson & Johnson, Citizenship & Sustainability Report, p. 9, 2015.

Citizenship and Sustainability 2020 Goals

Johnson and Johnson has put out public commitments for sustainability goals since the 1990s. New goals are developed every 5 years, and the current goals are called Citizenship & Sustainability 2020 Goals. The goals are set in three categories, *People* – which address aspects like making medicines available to those who can't afford it; *Practices* – that deal with suppliers improving their practices and having the healthiest employees in the world; and *Places* – reducing the environmental impacts of products and facilities. The Places goals cover various initiatives that reduce the impacts of facilities and products.

Places Goals

- Fully integrate sustainable design solutions into our product innovation processes
- Reduce our impacts on climate and water resources

Targets and Metrics

- New and existing products representing 20% of Johnson & Johnson revenue achieve Earthwards® recognition for sustainable innovation improvements.
- Increase recyclability of our consumer product packaging to 90+% in Key 1 Markets through design for recyclability and partnerships.
- Reduce absolute carbon emissions 20% by 2020 and 80% by 2050. Produce/procure 20% of electricity from renewable sources by 2020; aspire to power all facilities with clean/renewable energy by 2050.
- Conduct a comprehensive water-risk assessment at 100% of manufacturing/R&D locations and implement water-risk mitigation plans (WRMPs) at the high-risk sites.

Let's take a closer look at the product focus of the Places goal. There are two main objectives that address products in the Places category, recycling of consumer packaging and 20% of sales from Earthwards® products. Having goals to develop greener products are good, but they only address up to 20% of sales— what about the other products? Objectives have been set to demonstrate that all products are continually being greened for each business sector. All sectors have objectives to make packaging more sustainable by eliminating PVC, reducing packaging size, increasing its recyclability, increasing the use of post-recycled content, and increasing the use of bio-based materials. Each business sector has different objectives that will foster greener product improvements appropriate to their business units. Examples of some of the targets include sustainably sourcing palm oil, enacting deforestation programs, removal of target materials from products (e.g., toxic metals, PVC and brominated flame retardants), increasing the use of recycled content and bio-based materials in products, implementation of green chemistry methods, and providing end-of-life solutions to customers (e.g., electronic take-back programs).

Johnson & Johnson Greener Product Programs
• Earthwards® process is used to develop greener products using life-cycle thinking and seven key focus areas • Products making significant improvements receive the Earthwards® recognition • Sustainability Goals apply to all new products and packages to address enterprise-wide targets such as removal of PVC, use of PCR, and sustainable sourcing • Business sector appropriate initiatives have been set to green up processes and products

Johnson & Johnson uses the Earthwards® approach to make their products greener and uses life-cycle reviews to focus on the seven key areas for making individual product improvements (materials, waste, water, packaging, energy, innovation, and social). Environmental advances are also initiated by corporate-wide goals to green all products via business unit specific goals and all packaging by removing PVC, sourcing paper packaging sustainably, and incorporation of post-consumer recycled content.

Philips

Headquartered in Holland with 2015 sales of 24 billion euros, Philips has a robust greener product program, with 54% of sales representing more sustainable products. Having three main divisions, Lighting (household and industrial), Consumer Lifestyle (televisions, computers, vacuum cleaners), and Health care (CT scanners, ultrasound & diagnostic equipment), makes it challenging to have a unified greener products program. A good deal can be ascertained by studying this European-based conglomerates' approach to greener product design.

Philips has developed design guidance that can be applied to all business categories. Their objective is to offer more sustainable products to their customers through a design process focusing on a 10% environmental improvement in one or more of the following Green Focal Areas (compared to a predecessor product or competitor's product):

- Energy efficiency
- Hazardous substances
- Packaging

- Weight
- Recycling and disposal
- Lifetime reliability

Products that meet the 10% improvement are recognized with an eco-performance label. These products can be seen on the Philips website, indicated as *"green products"* (Philips 2015) (Figure 4.5).

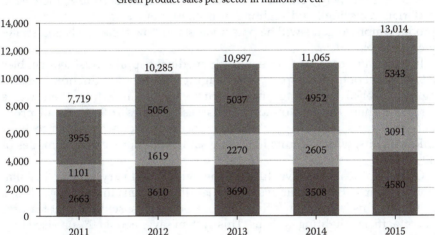

FIGURE 4.5
Philips Green Product sales 2011–2015.

The company reached record-breaking goals for Philips' green product sales in 2015 with 54% of all sales coming from their "green products," which is a 2% increase from 2014. In addition, investment in eco-innovation was set at 2 billion euros in 2010; Philips met this goal and then exceeded it in 2015 with an extra 495 million euros invested in eco-innovation (Philips 2015).

In addition to the greener products design program, sustainability goals have been set regularly since the 1990s. The newest sustainable goals called EcoVision have objectives, which apply to all of their operations.

EcoVision

- *Improving people's lives*—improve the lives of 3 billion people by 2020 through the use of their health-based products
- *Green product sales*—increase green product sales
- *Green innovation, including Circular Economy*—increase R&D spend in green innovation to >$2 billion

- *Green Operations*—environmental improvement of manufacturing facilities
- *Health and Safety*—protect worker health and safety
- *Supplier Sustainability*—adherence to supplier standards and encourage sustainability improvements (Philips 2015)

In order to make the most significant improvements, Philips has identified the key life-cycle aspects for each of their product categories: for Health care, it is reducing energy consumption, weight, and dose; Consumer Lifestyle, it is focusing on energy efficiency and closing material loops (e.g., increasing materials recycling), and Lighting, it is energy efficiency. Examples of the product improvements will help us understand how these goals and strategies are deployed.

In the health care area, a new PCMS medical supplies business enables a 90% reduction in air space in packaging and a 24% reduction in packaging material weight to support customers in reducing their waste streams. Their Home Monitoring business enables 76% re-use of products and parts. An energy-efficient cryocompressor was added to their MRI systems, which results in energy savings in non-scanning modes of 30–40%.

One of Philips newest ventures in the lighting industry was the installation of Philips LED lighting system in the Allianz Arena in Germany. Philips states that this is Germany's first and Europe's largest stadium that features energy-efficient lighting systems. This system saves about 60% on electricity and about 362 tons of CO_2 per year. Because of Philips' cloud-based ActiveSite platform, the stadium has lower maintenance and operating costs as well. As we can see, their "green products" are easily communicated to customers and have improved environmental performance comparison products (Philips Greener Products 2017).

Philips Greener Product Program Highlights

- Six Green Focal Areas to make products greener; energy efficiency, packaging, hazardous substances, weight, recycling and disposal, lifetime reliability
- Sustainability Goals focus on developing greener products
- Set enterprise-wide sustainability goals called Ecovision
- Focus on greening their supply chains

A Focus on Suppliers

Philips goes beyond greening their products and has set up programs to improve their suppliers' sustainability as well. The programs cover sustainability performance, management of regulated substances, conflict minerals, and other responsible sourcing initiatives. The Supplier Sustainability

Declaration (SSD) and Regulated Substances List (RSL) are documents that guide a suppliers' sustainability program. After suppliers agree to Philips' terms and conditions, auditors visit their sites to insure that they are 100% compliant. If the suppliers do not fully comply, then they cannot deliver to Philips.

Due to the nature of Philips business, the electronics industry requires the mining of various minerals. Philips has made sure not to source minerals that are associated with conflict minerals. A 2015 target was set to have 72% of their suppliers comply with their supplier sustainability program and the company exceeded this goal by onboarding 86% of their suppliers to the program (Phillips 2015). We are seeing more focus on the source of raw materials as part of a greener products story. This focus on suppliers and raw material sourcing is becoming a requirement to be able to call your product "greener."

Philips has established a comprehensive greener product development program. Use of focal areas and company-wide goals called Ecovision along with supplier objectives are cornerstones of their program. Having >50% of sales coming from greener products is a testimony to the effectiveness of their approach.

Samsung

The electronics industry has been an early adopter in developing greener products. The very nature of their products are a magnet for NGOs and ecologically aware consumers to put pressure on the manufacturers. The use of electricity and its link to CO_2 emissions, the short life cycle of some of their products, and past poor practices regarding disposal motivate electronic manufacturers to focus on developing greener products.

Taking a close look at the approach of a successful global electronics manufacturer headquartered in the Asia Pacific region yields some unique insights. Samsung Corporation, based in South Korea, is interesting to evaluate because of the diversity of electronic products it brings to the market. Some of the products include: televisions, cell phones, washing machines and dryers, computers, and printers.

Samsung participates in global sustainability initiatives as a member of the Electronic Industry Citizenship Coalition and has joined the Global e-Sustainability Initiative. They also conduct their business according to the United Nations Guiding Principles on Human Rights to ensure improvement on human rights conditions and the labor environment. Samsung has bridged the gap between social responsibility and business and endeavors to implement this in a meaningful way within the company (Samsung 2016a).

Samsung's approach to greener products started in 1995 when they adopted what they call the LCA approach. Environmental assessments have been required for NPD since 2004 when the "Eco Design Management Process" was established. More recently a new "Eco Design System (EDS)" and implementation of an Eco Rating System is employed. Samsung's plans for the future is to achieve 90% development rate of eco-products by 2020 as well as a total of 3.8 million tons of collected global waste of electronic products recycled by 2020 (Samsung 2016).

Samsung set a goal to develop sustainable innovative products that will reduce their products' impacts across the entire life-cycle. Their vision is "Providing a Green Experience, Creating a Sustainable Future," and their greener product development process is called PlanetFirst™. The Eco-Design process consists of two categories: Development Process and Eco-design activity. The Development Process includes: concept, plan, execution, and production. The Eco-design activity takes into account company and product targets, evaluation and improvement of the product, and final verification. Products are put into three categories based on the eco-grading scheme: Eco-product, Good Eco-product, Premium Eco-product. A goal has been set for 100% of products to achieve a minimum of "Good Eco-product."

Samsung started measuring these parameters and set goals to develop more energy-efficient electronics and reduce standby power consumption. Achieving this objective will not only benefit the environment but also reduce operational costs for customers. They also have adopted a circular resource management system to take into consideration the total environmental aspects in the product design. Every eco-product goes through a rigorous LCA of the following stages: supplier purchases, design and production, distribution logistics and packaging, use and reuse of the product.

An example of their work toward this goal is the switching technology for televisions from LCD to the lower impact LED. The TV (UE60J6150) uses less power in normal operation and in standby mode than a similar LCD product. The TV has various features including an EU Energy label of A++, light intensity sensors, energy-saving mode, auto power down function, and reduction of product weight by 27% (vs. UE58H5270AS).

A significant focus area for electronics is minimizing the use of the hazardous ingredients heavy metals and brominated flame-retardants. Samsung implemented a supplier program to address this issue called the Eco-Partner Certification Scheme in 2004. The company requires the Eco-Partner Certification as well as compliance with ISO14000 and OHSAS18000 certifications. The object of this initiative is to reduce the use of hazardous parts and raw materials, and assist suppliers in setting up environmental quality systems. If a supplier puts a process in place to manage the substances, Samsung issues a certification of compliance to the standards. Additional steps are taken by inspecting components and raw materials to prevent the use of hazardous materials (Samsung 2016b).

Eco-Labels

Samsung pursues third-party certificates to demonstrate their improvements. As of 2015, they obtained certifications in 11 different countries. Some of these are USA (EPEAT), Germany (Blue Angel), Sweden (TCO), EU (Eco Flower), Scandinavia (Nordic Swan), and Canada (Environmental Choice). An impressive 2,218 models in various product groups (printer, PC, monitor, TV, DVD player, refrigerator, and washing machine) have received environmental certifications. This is a very strong commitment that demonstrates to customers that significant product improvements were made.

Further examples of product improvements are demonstrated through recognition by the US magazine *Good Housekeeping* for Samsung's front-loading washing machine as the "best washer" for using the least amount of water among tested washers. Another innovation within the washing machine category was the "Eco Bubble" washing machine technology which reduces washing time and is 40 times faster compared to conventional front-loading washers. The washing machines are able to provide the same thorough cleaning *without using hot water* which is the biggest life-cycle impact for a washing machine and a big energy saving technique (Samsung Eco Bubble 2017).

Samsung's Eco-Product initiatives

- Use the Eco-Design Management Process to make greener products
- Goals have been set to increase the amount of "good" and "premium" eco-products
- Eco-labels are pursued with over 2,218 models receiving third-party environmental certifications
- Initiated voluntary take-back programs: 54,354 tons of waste electronics were recycled in 2015

Other examples of eco-innovative products include the mobile phone. The Galaxy S6 Edge comes with a high-efficiency charger (charging efficiency of 82% and standby power of 0.02 W), ultra-power saving mode, recycled plastic used for the charger (20%), and 100% recycled paper packaging. The (LS27E65UDS) computer monitor has an annual power consumption reduced by 36% (compared to LS27C65UDS), recycled plastic used (30%), and an Intertek Greenleaf certification (Samsung 2016). All of these products are good examples of what would delight consumers, because these improvements just don't improve the environment, they save the customers' money.

End-of-Life Management

End-of-life issues are important to consider for any product stewardship program. This is especially true for electronics. Poor disposal practices resulting in environmental and human health problems have been well publicized. Keeping this in mind, Samsung has voluntarily initiated take-back programs for waste electronics. They set up Korea's first waste electronic product recovery and recycling system in 1995. The program is intended to prevent the illegal disposal of electronics through incineration and land filling and to encourage the recovery of valuable materials.

Expanding on this concept, global take-back programs have been put in place in countries where electronics recycling is not mandatory, like the United States. An estimated 54,354 tons of waste electronics have been taken out of disposal systems and put into recycling systems as a result of this program in 2015. Samsung has cumulatively collected 2.26 million tons of waste products from 2009 to 2015 and has the goal of collecting a cumulative total of 3.8 million tons by the year 2020 (Samsung 2016).

Samsung has emphasized bringing more eco-products to the market. Through the use of LCAs, efforts have been focused on minimizing the most important impact areas, energy use, and standby mode. Improvements are further encouraged with new product design checklists and guides. A scoring system called the eco-rating system ranks products' greenness. Their objective is to bring to market more good and premium rated eco-products.

Apple Inc.

One of the most innovative companies in the world is the electronics manufacturer Apple, the creator of the Macintosh computer, the worlds' thinnest laptop, the extremely popular iPhone, the functional and easy-to-carry iPad, and one of their newest technologies, the Apple Watch. Their innovation has been applied to their sustainability program as well.

Apple had been criticized by environmental groups for not being proactive enough on making their products greener; in 2007, Greenpeace ranked Apple last among major electronics manufacturers on environmental issues. They claimed that the use of hazardous materials like PVC and toxic flame retardants was lagging compared to similar electronic product manufacturers. Since the generation of this report, there has been a concerted effort to make significant product improvements. Eventually, Greenpeace acknowledged that Apple was moving in the right direction. They concluded that their campaign highlighted Apple's shortcomings to their customers, which in turn caused positive changes in their products. Further, they believed that their public reporting of Apple's performance forced Apple to become a green leader (Greenpeace 2010). Regardless of the reasons why Apple has made strides to bring greener products to market, they do present their progress, as you would expect, in an innovative way.

Apple has set three focus areas to make the most impact possible. Their first priority is to reduce Apple's impact on climate change through use of renewable energy sources and energy-efficient products and facilities. Their next priority is to conserve resources, and their third priority is to use safer materials in all products and processes (Apple 2016).

Use of Life-Cycle Assessment

Apple uses LCA to measure its environmental footprint. The key indicator used for determining their footprint is greenhouse gas emissions. For 2015, Apple estimates that it emitted 38.4 million metric tons of greenhouse gas emissions. Emissions are calculated from manufacturing, transportation, use, and recycling of products. Their analysis determined that a majority of their emissions come from product manufacturing (77%) and customer use (17%). Therefore, to improve their footprint, they decided to focus on sourcing lower carbon materials, partnering with suppliers to add clean energy to their facilities and produce clean energy at Apple offices, retail stores, and data centers globally, as well as adjust recycling and shipping strategies.

To address the use of materials in their products, a design for the environment concept, dematerialization, was employed very successfully. Looking at their product lines, it's evident that design engineers have fostered the innovation of thinner, smaller products that deliver the same or more benefits as competitors. The reduction of raw materials used in electronics yields significant environmental improvements. Consider the impact benefits from fewer metals having to be mined and transported to manufacturing sites, lighter products which require less energy and chemicals to process and less transportation emissions are generated since more packages can fit on trucks or planes during shipping.

An example of dematerialization combined with enhanced performance is the 21.5-inch iMac. This computer is faster, has more features, and has a larger screen than the first-generation 15-inch iMac. Amazingly, it uses 97% less power in the sleep mode and the packaging uses 35% less material.

Apple's Eco-Design Focus Areas

1. Use less material
2. Ship with smaller packaging
3. Be free of toxic substances
4. Be as energy efficient and recyclable as possible

Removing Toxic Materials

Raw materials needed for electronic products contain toxic materials. Since Apple received a good deal of criticism for not being proactive in removing toxic materials from their products, they decided to be more public about their accomplishments and future goals. Apple has reached their goal to lead their industry in reducing or eliminating environmentally harmful substances such as toxic metals, brominated flame retardants (BFRs), phthalates, and polyvinyl chloride (PVC).

A typical cathode ray tube (CRT) contains an amazing amount of lead (approximately 3 pounds or 1.36 kg). Apple became the first company in the computer industry to eliminate CRTs back in 2006. At the time, all the leading manufacturers of PCs were still using CRTs (Apple Inc. 2010). Apple claims that they are far ahead of their competitors in reducing toxic materials. Every product is free of brominated flame retardants (BFRs), elemental bromine, and chlorine and every display has mercury- and arsenic-free glass.

Continuing their leadership in toxic material removal, Apple completely removed all PVC from their power cords and headphone cables. This four-year process eventually led to the use of safer materials. Apple found the perfect blend of durability, safety, and environmental performance. They have replaced all PVC with non-chlorinated and non-brominated thermoplastic elastomers.

Packaging

Another objective to improve product environmental performance is to minimize product packaging. Apple's efforts have a double benefit—less packaging materials and reduced greenhouse gas emissions during product transportation. Overall, 99% of Apple's product packaging came from recycled paper or from forests that have been sustainably sourced. Apple focuses on using paper efficiently, sourcing virgin paper responsibly, and protecting sustainable forests. As an example "U.S. retail packaging of iPhone 6s is 20% lighter and consumes 34% less volume than the first-generation iPhone packaging" (Apple 2016).

Energy Efficiency

Focusing on designing more energy-efficient products is one of the most critical aspects of eco-design. The new Mac minicomputer is the smallest desktop thus far, measuring just 6.5 inches square. The mini is reported to be the world's most energy-efficient desktop computer. It uses less power than a single 13-watt CFL light bulb! If you take these savings across the hundreds of millions of computers sold in the world, you can imagine the tremendous amount of energy savings that can be achieved (Sillydog 2017). This little but powerful computer exceeds the Energy Star requirements sevenfold (Apple 2016).

Having third-party standards to verify claims is helpful to build credibility. Energy efficiency claims have been confirmed by meeting or dramatically exceeding the U.S. EPA's Energy Star criteria. Making products that last longer is a design-for-environment principal. A longer lasting product reduces the resources needed to make new ones. When developing a new computer, Apple tests the MacBook keyboard's durability by applying millions of clicks to each button. This ensures that the product will last a long time. All Apple products are built to stand the test of time, with regular software updates to keep products up to date and reduce the need for replacements (Apple 2016).

Recycling Initiatives

No matter how long a product is designed to last, it has an end. In order to lower their carbon footprint, Apple has placed a priority in recycling products in the region where they're collected to reduce harmful carbon emissions. Electronic products have considerable concerns because of its metallic components and flame retardants. Many electronic manufacturers have established end-of-life management programs to address this concern. Apple has established recycling initiatives in 95% of the countries where their products are sold. Their programs have enabled the recycling of more than 597 million pounds of products since 1994. In 2015 alone, Apple's recycling program collected 90 million pounds of e-waste.

Part of this initiative was to insure that recycling partners are acting responsibly; therefore, auditing programs and guidelines are critical. Apple does not permit firms to dispose of electronic waste in solidwaste landfills nor permits them to be incinerated; this is not always a guarantee in developing nations. Enabling the take-back and recycling of products helps to bolster a company's green image and is a strong statement to customers, showing that environmental issues are front of mind when a product needs to be disposed of.

To address this issue further, a line of robots that can disassemble iPhones, called Liam, was introduced in 2015. Liam can dissemble iPhones at a rate of one iPhone every 11 seconds. These robots separate and sort high-quality components that can be recycled. Through the use of Liam, Apple has diverted more than 89 million pounds of e-waste from landfills (Apple 2016).

Product Environmental Report

Several companies publish product-specific environmental reports, which I think is really a good idea because it's a single source of truth for all the environmental information a customer or stakeholder would want for a product. Apple has posted several reports on their website. Since their latest innovative product, the Apple Watch, a review of how Apple evaluates this product will help in understanding their approach to product greening and demonstrate a best practice for communicating a product's environmental footprint.

The categories that are cataloged for each product include:

1. Climate Change
2. Energy Efficiency
3. Material Efficiency
4. Packaging
5. Restricted Substances
6. Recycling

The report details the Apple Watch Series 2 environmental impacts with graphs and tables. The life-cycle greenhouse gas emissions are estimated at 30 kg CO_2e. The emissions from the greatest to least are reported as 68% production, 19% customer use, 10% transportation, and 3% recycling. Apple Watch Series 2 is reported to use energy-efficient components such as a Power Adapter that is more efficient than the U.S. EPA Energy Star specifications for power supplies. Power consumption is reported on for the sleep, idle, and adaptor modes.

Apple strives to use recyclable materials like aluminum and glass. They are leading the industry in ultra-compact product and packaging designs which reduce the overall material footprint of a product. Materials used are charted, and the weight of various components are cataloged: plastic (41 g), steel (12 g), aluminum (8 g), circuit boards (4 g), glass (3 g), ceramic (3 g), display (1 g), magnets (1 g), and other (1 g).

Packaging is addressed by the type and weight of material for the retail box and shipping box, for example, paper (corrugate, molded fiber), polystyrene plastic, and other plastics. The Apple Watch Series 2 retail box contains at least 39% recycled content. The aluminum case for the watch is made from fiber-based materials originating from recycled content, agricultural by-products, or sustainably managed sources. Apple claims that its packaging is extremely material efficient, and it is obvious from the photos in their product's environmental report that they are designed to the form of the product with little excess (Figure 4.6).

To show progress toward the use of restricted substances, the Apple Watch Series 2 complies with the European Directive on the Restriction of the Use of Certain Hazardous Substances in Electrical and Electronic Equipment (RoHS Directive). This Directive covers lead, mercury, cadmium, hexavalent chromium, and PBB and PBDE brominated flame retardants (BFRs). Apple products also comply with the European Regulation on Registration, Evaluation, Authorization and Restriction of Chemicals (REACH). Apple reports that they have reduced the following hazardous ingredients even

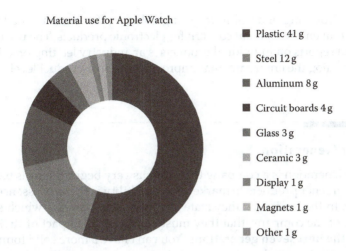

Material use for Apple Watch

- Plastic 41 g
- Steel 12 g
- Aluminum 8 g
- Circuit boards 4 g
- Glass 3 g
- Ceramic 3 g
- Display 1 g
- Magnets 1 g
- Other 1 g

FIGURE 4.6
Apple Watch material use. (From Apple Watch Environmental Report 2017.)

further than regulatory requirements: mercury, brominated flame retardants (BFRs), PVC, and beryllium.

Recycling is reported in a general manner in the product report. Use of less raw materials and inclusion of recyclable materials in the product design are emphasized; the result is less waste at end-of-life, plus maximum recyclability. There are take-back and recycling programs in 99% of the regions where the products are sold (Apple Watch 2017).

In addition to the very detailed environmental report, a status report is maintained to highlight some key features.

Apple Watch Series 2 Environmental Status Report

Apple Watch Series 2 is designed with the following features to reduce environmental impact:

- Mercury-free
- Brominated flame retardant-free
- PVC-free beryllium-free
- Complies with European REACH regulation on nickel
- Retail packaging contains at least 39% recycled content

Apple is a leader in innovative products and they have taken an innovative approach to product stewardship. One of their most significant accomplishments is the light sizing of electronic devices. They have been a leader in making some of the thinnest laptops and smart phones in the electronic industry. This initiative alone has gone a long way to advance greener products. Strides have also been made in reducing toxic materials and designing devices for longevity. Apple has some of the longest lasting batteries in the electronics industry. Providing end-of-life recycling solutions for customers addresses an important environmental concern for electronic products. The use of environmental reports on individual products is an industry leading practice that helps to catalog the environmental improvements on a product level.

Seventh Generation

Seventh Generation is a company that from its very beginning was based on bringing greener products to market. Sustainability is the very essence of the company. In their mission, they state an Iroquois Indian law, which shaped the name of the company, that they must **consider the impact of their decisions on the next seven generations**. You can't have a more solid foundation for a company to base their greener product program on. So it makes sense to evaluate the approach of Seventh Generation to understand how a company with green at its very roots develops products (Seventh Generation 2016).

Seventh Generation makes various home use products such as botanical disinfectants, dishwashing detergents and soaps, hand soap, laundry detergents, surface cleaners, household paper and supplies, baby diapers and wipes, baby laundry, organic tampons and pads. Their product developers, while focusing on effectiveness, are to take into account economic, environmental, and health factors. A product scorecard is used to assist in developing greener products.

Setting Sustainability Goals

The use of sustainability goals to drive product improvements is part of Seventh Generation's strategy. Objectives have been set to reduce impacts like greenhouse gas emissions and increase the use of plant-derived ingredients. One unique approach is that they have set an internal price on carbon at $6/ton. They hope this will encourage purchasing groups to buy materials that have lower carbon associated with it.

2020 Nurture Nature Sustainability Goals

DECREASE OUR CARBON FOOTPRINT
- All energy from non-fossil sources
- All clothes washing in cold water

CHOOSE PLANTS NOT PETROLEUM
- All products and packaging bio-based or recycled

PRODUCE ZERO WASTE
- All products and packaging recyclable or biodegradable

SOURCE SUSTAINABLY
- All agricultural ingredients certified sustainable

Packaging

One of the most visible aspects of a consumer product companies' sustainability is packaging. Realizing this, Seventh Generation has made a conscience effort to place greener packaging on store shelves. Significant progress was made on the eco-friendliness of the type of material used and use of recycled content.

An analysis of the type of packaging used resulted in a switch from polyethylene terephthalate (PET) to a high-density polyethylene (HDPE) because it takes less energy to make. In addition to this, a move to incorporate more post-consumer recycled content (PCR) into each package was made. Also, since HDPE is used in the common milk jug, Seventh Generation is using these discarded containers to manufacture their *dishwashing liquid bottles*.

Setting high expectations for your organization is a necessary step in developing greener products. This includes setting stretch goals such as striving for 100% PCR content in bottles before anyone else in your category. Leading isn't easy; "failed attempts" due to PCR being more rigid created problems; however, working with suppliers, progress was made and now the hand dishwashing liquid, fabric softener, and bleach containers use 90% PCR.

Post-consumer content packaging has increased over time, and Seventh Generation has also emphasized the use of renewable materials. Packaging goals for 2020 strive for all packaging to be bio-based or recycled, biodegradable or recyclable. In 2015, Seventh Generation reduced their use of virgin plastic derived from petroleum by 195 metric tons. They have also increased the recyclability of their packaging by 20 metric tons. The materials that are used for plastic packaging follow this hierarchy: post-consumer recycled plastic, virgin plant-based plastic, and virgin petroleum-based plastic. Seventh Generation's 100 oz. 2x liquid laundry detergents use 80% post-consumer recycled plastic and 20% virgin plastic. The virgin plastic used in the laundry detergent bottles was changed from petroleum-based plastic to plant-based plastic made from sugarcane and is also 100% recyclable (Seventh Generation 2016).

Materials

Having a natural products line requires a greater focus on the ingredients that go into a product. Therefore, relentless effort to improve your formulas is critical to maintaining your customers trust. An example of the continual greening of products is the 27% increase of renewable materials in dishwashing liquid accomplished in 2009. They also removed an undesirable ingredient that was part of surfactant 1, 4-dioxane, a possible human carcinogen.

Future directions for greening products are guided by Seventh Generation's new sustainability goals: all products and packaging create zero waste and are biodegradable or recyclable by 2020; all ingredients and materials are biobased or recyclable by 2020.

Another ingredient activity that demonstrates attention to detail is the commitment to sustainably sourcing materials containing or derived from palm oil. Palm oil has become a significant issue because it is sourced in regions where portions of rainforests have been cut down to establish palm groves. To address this issue, a goal was established that 100% of palm oil will be sustainably sourced.

Unique ingredients have been used to offer customers products with more natural materials. New disinfectant cleaners were developed using thymol, a component of thyme oil derived from the garden herb thyme. Seventh Generation claims that these cleaners kill 99.99% of germs. A review of the ingredient list from their Disinfecting Multi-Surface Cleaner demonstrates a commitment to natural ingredients. It reads like a label on a salad dressing bottle.

Disinfecting Multi-Surface Cleaner Ingredient List

Aqua (water), thymol (component of thyme oil), sodium lauryl sulfate (palm kernel or coconut-derived cleaning agent), copper sulfate pentahydrate (bluestone) (mineral-derived water mineralizer), citric acid and sodium citrate (cornstarch-derived water softeners), essential oils: *Origanum vulgare* (oregano) oil, *Cymbopogon nardus* (citronella) oil, *Cymbopogon schoenanthus* (lemongrass) oil (Seventh Generation 2017).

Transparency is another key aspect to Seventh Generation's commitment to greening their product ingredients. They commit to disclose all ingredients in their products and do so on their website and on product labels. A visit to their website will enable the viewer to see a list of everything that is in a Seventh Generation product. The list of ingredients above for their Multi-Surface Disinfectant Cleaner is from their website. There is also a glossary that explains terms used to describe ingredients used like synthetic, plant derived, plant modified, etc. This is a best practice because it makes it easy for customers to understand what the purposes of ingredients are and how they are being derived.

Key Aspects to Seventh Generation's Program

1. The company is based on having greener products: They consider the impact of their decisions on the next seven generations
2. Big emphasis on natural ingredients & incorporation of PCR in packaging and recyclability
3. Set sustainability goals to encourage greening products
4. Commitment to transparency of all product ingredients

Seventh Generation is a company whose very essence is based on bringing greener products to the marketplace. It is obvious that they are fully committed to this concept when reading their sustainability report. There is a big emphasis on natural ingredients and the use of PCR in their packaging. A further focus on sourcing ingredients in a sustainable manner and a transparency commitment to listing all ingredients for each product sold rounds out the key elements to their product greening strategies.

Method

You need not go any further than the company tagline "people against dirty" to understand Method's commitment to sustainable products. The company was formed in 2001 with the goal of providing natural cleaning products. Looking at their website and a photo of their very youthful looking founders, you know you're dealing with a cutting-edge, innovative company. Method makes baby products (shampoos and lotions), hand cleaners, body wash, laundry detergent, and household cleaners. Their goal is "making products safe for every surface, especially earth's" (Method 2016).

Of all the companies evaluated in this book, Method has the most comprehensive approach to greener product design. According to company statements, they believe that business can be an agent for positive social and environmental change. Method was one of the first companies in 2007 to become a B corporation. The vision of becoming a B corporation "is simple, yet ambitious: to create a new sector of the economy that uses the power of business to solve social and environmental problems" (Method 2016).

A comprehensive list of criteria is used when designing products for "true sustainability." Method rightly states that focus cannot be on just one element like carbon footprint or safety; new products require a holistic evaluation. Method explains that only focusing on one aspect to green a product may result in less than optimum results. For example, natural should not be the only dimension that is focused on for a product. Natural ingredients are important, but some natural ingredients can be toxic. Method uses natural as the place to begin and also evaluates all other aspects of a product to insure it's making the most sustainable decisions possible.

Packaging

Sustainable product assessments must also include consideration of the product packaging. An important consideration for packaging is the recyclability of the materials being used, or as Method states, using "bottles made from bottles." The use of PCR is paramount to designing sustainable packaging according to Method; they state that they only use 100% recycled plastic to make their bottles, which has a 70% lower carbon footprint than using products that are made of virgin plastic. They also strive to reduce the mass of the plastic in each bottle, as evident by the marketing of pouches for some products which they claim to reduce 78–82% water, energy + plastic savings versus a bottle. The company also focuses on using plastics 1 and 2 which are the most highly recycled resins. Another way that Method focuses on creating more environmentally conscious packaging is by using refill pouches for cleaning products as opposed to buying a new bottle. The refill pouches save about 80% of plastic, water, and energy as opposed to a bottle. Lastly, Method boasts that they are the first company to make bottles made of ocean plastic and post-consumer recycled plastic (Method 2016).

Cradle to Cradle Design

Method says that they follow the precautionary principle regarding ingredients: "if there's a chance an ingredient isn't safe, we don't use it. Period." (Method 2016). A way to demonstrate this commitment is embracing the Cradle to Cradle design theory. This theory was developed by McDonough Braungart Design Chemistry (MBDC), a sustainability-consulting firm. This firm also issues certifications for products that follow their Cradle to Cradle® Framework. The certification is issued on a product-by-product basis and covers material health, material reutilization, water, energy, and social fairness (MBDC 2016).

Cradle to Cradle (C2C) design encourages following natural cycles, or as MBDC says, "nature's biological metabolism." The main idea behind C2C is to design products and materials with life cycles that are safe for human health and the environment and that can be reused in a closed-loop fashion (MBDC 2016).

Key Business Benefits of Cradle to Cradle

- Evaluates products
- Reduces risks
- Reduces costs
- Third-party independent evaluation and certification

The C2C design framework has been included in all the products that Method designs. In 2016, they had over 100+ products that were C2C certified, one of the highest number of any company. Products such as handwash, dish detergent, laundry detergent, and body wash are certified. It is Method's goal to have their entire product line achieve certification. The C2C criteria are very comprehensive. There are five categories in which a product is evaluated. In each area, there are criteria set that leads to achievement of silver, gold, and platinum levels.

C2C Categories of Criteria for Certification

- Material Health
- Material Reutilization
- Water Use
- Energy Use
- Social Fairness

The C2C products program assesses the environmental effects, hazards, and risks of the product that is being evaluated by looking at a complete materials list of the product and the product packaging. As an example of how products are evaluated, the Material Reutilization category "rewards products that contain recycled or renewable materials." Manufacturers are expected to develop and implement a strategy to close the loop on the product at the end of its useful life, at the highest levels of certification (MBDC 2016).

Formulation Approach

The following five steps are used when developing products:

- *The Precautionary Principle*—if there's a chance an ingredient isn't safe, we don't use it.
- *The Dirty Ingredient List*—ingredients that many others use, but we don't.
- *The Highest Standard*—creating cradle to cradle products.
- *Comprehensive, Third-Party Assessment*—ingredient reviews are completed by independent researchers using scientifically peer-reviewed materials.
- *Smart Science*—Green chemistry techniques ensure effective and safe formulas.
- *External Validation*—Cradle to Cradle certifications.

Having a robust process is critical to bringing greener products to market. This includes having principles and procedures that product developers can use. As indicated above, Method uses a comprehensive evaluation process to bring greener products to market, giving product developer's guidance about which materials to steer away from is an important element. A list of undesirable or "dirty ingredients" is maintained which formulators

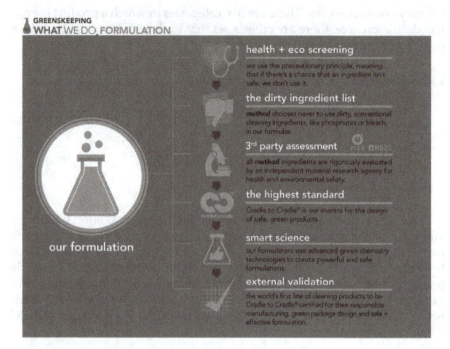

FIGURE 4.7
Methods formulation approach.

must avoid. The list includes materials such as chlorine bleach, triclosan, EDTA, phosphates, 2-butoxyethanol, phthalates, paraben, ammonia, and MEA (ethanolamine) (Method 2016) (Figure 4.7).

To further bolster trust in their brand, each product page on their website lists every ingredient and fragrance used. Other companies have committed to this form of transparency; however, Method goes further and *makes available an evaluation of each ingredient by a third party*. Product labels have a list of the ingredient names and a link to where you can get more information online.

Key Aspects to Method's Program

1. The company mission is founded on greener products: "People against dirty"
2. Holistic greener product design process & guides
3. Commitment to Cradle to Cradle certification
4. Third-party assessment of all products and ingredients
5. All product ingredients are made publicly available
6. Emphasis on the use of PCR in packages

Method has been a deep green company from its very beginning. All of their products are based on greener product design principals. They consider an extreme amount of details when developing products from the efficacy and toxicity of raw materials, avoiding the most problematic (or dirty) chemicals, to the amount of PCR in packages. All products undergo a third-party review to insure that they meet the company's goals for greenness. Method has committed to getting all products certified by C2C. All of their product ingredients and fragrances are available on their website. The company believes that they can make a difference with their products and are committed to bringing healthier and environmentally safe products to market.

Proctor & Gamble

Proctor & Gamble (P&G) is a company with over 170 years in business and, with $65.3 billion in 2016 sales, is the world's largest consumer package goods company. They have iconic brands which are sold globally. Some of their key brands are Tide laundry detergent, Luvs disposable diapers, Gillette razors, Olay beauty products, and Duracell batteries. Being a large multinational company with over $65 billion in sales (2016), it is interesting to see how they have made their sustainability programs and initiatives very public and an obvious enhancement to the company's equity. A big part of their sustainability initiative is their product greening efforts (P&G 2016)

The Use of Goals to Improve Performance

P&G received significant publicity when they announced their long-term environmental vision. Of these long-term goals, two of them focus on their products, which demonstrate that they understand that their biggest environmental and social impacts come from manufactured goods.

P&G's Environmental Vision Has Four Very Big Objectives

- 100% renewable energy will power our plants
- 100% renewable or recycled materials will be used for all products and packaging
- Zero consumer and manufacturing waste will go to landfill
- Designing products that delight consumers while maximizing the conservation of resources

Having bold goals is good, but it's necessary to set interim targets to demonstrate progress that will put the company on a path to realizing the bigger objectives. P&G has set 2020 targets as stepping stones toward their long-term vision. Progress has been made on all of their long-term objectives as of year end 2015. Let's focus on the progress they reported from 2010 to 2015 associated with their products; it's a good list of positive actions a company can take to make meaningful improvements to their products.

Third-party certifications were in place for 100% of virgin wood fibers in all hygiene products and 61% of wash loads are in low-energy cycles (cold water and high-efficiency low-energy clothes washing). The company implemented palm oil commitments on traceability and supply-chain engagement, as well as created technologies to replace petroleum-derived plastics and cleaning agents with renewable materials. Paper packaging (86%) contains either recycled or third-party-certified virgin content. Paper packaging reduced packaging by 12.5% per consumer use and increased the use of recycled resin in plastic packaging by 30%; 450 million people have access to water-efficient products (P&G 2016).

Product Innovation Using a Life-Cycle Approach

To reduce product environmental impact and spur product innovation, P&G uses an LCA approach. Looking at a product's full life, from raw materials to manufacturing and product use, helps identify the most important areas to focus on to make the greatest reduction in environmental impact. Prior to consideration of life-cycle impacts, companies focused primarily on reducing the footprint at the manufacturing facility. Environmental initiatives that were focused on reducing energy, water use, and waste generation all seem like the right thing to do.

Perhaps one of the best demonstrations of the application of LCA is cold water laundry detergent. If you think of P&G's reach and scale, they have tremendous ability to reduce environmental impacts. As an example, Ariel and Tide cold water detergents enable consumers to save energy by turning down the temperature dial on every wash, using up to 50% less energy per load than warm washes (P&G 2017).

Proctor & Gamble evaluated the full life cycle of using a mop and bucket for cleaning and realized that consumers waste massive amounts of water by cleaning their houses with a traditional mop and bucket. This enabled product developers to focus their efforts on the most important area and generated a significant sustainability innovation. The Swiffer Wet Pad is a more-convenient way to clean that allows the consumer to combine a mop, bucket, and cleaning solution. A household that uses this product will save more than 70 gallons of water every year in comparison to a mop and bucket—more convenient and lower environmental impact.

Another example of the LCA approach is the evaluation of Pampers disposable diapers. The review indicated that the sourcing and production of raw materials is the biggest impact area. Development of a new technology that uses less absorbent materials resulted in a thinner diaper; 16% less raw materials are used in Pampers size 4 diapers compared to the previous diapers. This technology means that just in Europe, impacts will be lessened by 180,000 tons of material, 11% less energy, and reduce the number of pallets needed which in turn lessens the amount of trucks on the road. The use of fewer materials addresses the biggest life-cycle impact.

Proctor & Gamble's Greener Products Program

Long-Term Goals

1. Zero consumer waste will go to landfill
2. Designing products that delight consumers while maximizing the conservation of resources
3. Use LCAs to work on the most important product impacts

P&G is the largest package good consumer product company in the world. Marketing products that are prominent in most households gives P&G a tremendous opportunity to make big impacts with greener products. Very bold goals have been set demonstrating significant commitment to reducing the impacts of their products. The use of LCA guides product developers to focus on the most important areas needing product improvement, which have led to innovations. P&G sets a high bar for consumer product manufacturers to follow and indicates that there is a strong pull for greener products in the marketplace.

Unilever

Another large consumer's product manufacturing company which has made significant sustainability commitments is Unilever. Headquartered in London, Unilever states that on any given day, two billion people use their products. With sales of 53.3 billion euros (2015), they manufacture a lot of food products, refreshment categories, home cleaning, and personal care products. They have over 400 brands, many of which are household names, such as Lipton, Knor, Axe, Hellmann's, Heartbrand Ice Creams, Magnum, Rama, Ben & Jerry's, Rexona, Sunsilk, and many more (Unilever 2015). When you manufacture products used by 2 billion people every day, you have a big opportunity to impact the world.

OUR VISION is to grow our business, whilst decoupling our environmental footprint from our growth and increasing our positive social impact. (Unilever 2017)

The company's vision embraces sustainability. To fully understand how this initiative impacts product design, we will evaluate some of the accomplishments and direction Unilever has taken.

Unilever gained a lot of notoriety by making a bold commitment to a long-term sustainability initiative in 2010 called the Sustainable Living Plan. Their CEO is very apt to speak about this initiative as a major undertaking for their company and positioned it as a strength of their brand. The plan embraces all aspects of sustainability and ties them into sourcing and selling their products. The idea is to increase sales in a sustainable manner—to enable billions of people to increase their quality of life, without increasing their environmental impact.

Sustainable Living Plan

The Sustainable Living Plan has three categories of focus:

- Improving health and well-being
- Enhancing livelihoods
- Reducing environmental impact

Improving Health

Unilever wishes to help more than 1 billion people take action to improve their health and well-being by 2020. They plan to do this through their products. Progress is reported by demonstrating the impact of their products as stated in their 2015 Sustainable Living Plan Progress Report: "around 482 million people reached by end 2015 through our programs on hand washing, oral health, self-esteem and safe drinking water, 337 million people with Lifebuoy; 71 million with our toothpaste brands; 19.4 million through Dove self-esteem programs; and 55 million with safe drinking water from Pureit. Pureit also provided 78 billion liters of safe drinking water by the end of 2015" (Unilever 2017).

Enhancing Livelihoods

Unilever plans to enhance the livelihoods of millions of people as they grow their business. There are three key areas: fairness in the workplace, opportunities for women, and inclusive business. The focus of this objective is more on social issues such as responsible sourcing, fair compensation, encouraging more females in management, enabling women to run businesses, and help smallholder farmers.

Reducing Environmental Impact

A big goal was set to reduce environmental impact by *halving the environmental footprint* of the making and use of products as they grow their business, by 2030. Taking steps toward this objective, Unilever committed to becoming **'carbon positive'** by 2030 which consists of 100% of our energy from renewable sources.

The plan to reduce their emissions begins with a review of the companies' entire greenhouse gas life cycle and focuses on the higher impact areas. Company-wide emissions were calculated to be around 59 million tons of CO_2. As seen in the Figure 4.8, manufacturing and distribution represent less than 5% of the total GHG footprint, and consumer use is over 60%. The product categories that make the largest contribution to the GHG footprint are those that require consumer use of heated water like showering, washing hair, and laundry.

So how does a company reduce emissions not under their control? Unilever does it by encouraging consumers to reduce their greenhouse gas emissions while washing and showering such as encouraging consumers to adopt better laundry behavior—habits such as correct dosing, lower-temperature washing, washing a full load, and using shorter wash cycles. These habits are communicated to customers on pack, in-store, and online. Some washing machines have the ability to run at lower temperatures and shorter wash cycles, messaging to encourage these lower impact settings is part of the plan. There are also new products that help to change habits, such as dry shampoos or 'quick wash' laundry detergents (Unilever 2017).

An important environmental aspect for a company that sells food products is responsible sourcing initiatives. These initiatives are becoming a greater

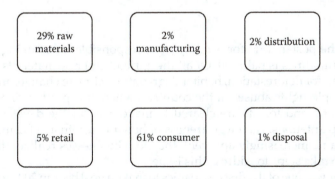

FIGURE 4.8
Unilever Greenhouse Gas Footprint 2014–2015. (Unilever Sustainable Living Plan, Summary of Progress 2015. p. 5.)

area of focus for companies' greener product development programs. Let's take a deep-dive look at the plan, which I believe is one of the most comprehensive and an industry's best practice.

Sustainable Sourcing

Working with suppliers can make a much bigger impact than focusing on a company's own manufacturing operations. Many of Unilever's products depend on agriculturally derived raw materials. Some developing countries have depleted important biological areas like tropical rain forests to provide agricultural products. NGOs have urged focusing on sustainable production methods to combat this problem.

Unilever has made it a goal to achieve 100% of their agricultural products from raw materials that are sustainably sourced.

Unilever has made it a goal to achieve 100% of their agricultural products from raw materials that are sustainably sourced. They created a program called the Unilever Sustainable Agriculture Code, a 72-page document of guidelines on how to sustainably farm key crops, which all of their suppliers must comply to. In addition to this Code, there is a "Responsible Sourcing Policy" which also has mandatory requirements for suppliers like complying with all laws, treating workers equitably, and workers are to be of the appropriate age and paid fairly, land rights of communities and indigenous people are protected and operate in a manner that embraces sustainability and protects the environment.

As a large purchaser of palm oil and the largest purchaser of tea in the world, two of the most depleting agricultural raw materials, Unilever is working toward being a leader in sustainable sourcing. The company is committed to sustainably sourcing these two key commodities.

Palm Oil

Perhaps the poster child for needing to be responsible for sourcing agricultural commodities is palm oil. Of all the agricultural raw materials, palm is most noted for deforestation, habitat degradation, climate change, and indigenous people rights abuses in the countries where it is produced, primarily because land and forests are cleared to make way for the development of palm oil plantations. As the awareness of how unsustainable farming palm oil is became increasingly apparent, the more businesses realized that they needed to take steps to address this issue.

Unilever was one of the first companies to make a goal back in 2009 to commit to sustainably sourcing 100% of their palm oil. The company reached this goal as of 2015, all palm oil has been sourced sustainably. They were able to reach their goal 3 years ahead of their plan due to the development of GreenPalm certificates. Growers can earn GreenPalm certificates from the Roundtable on Sustainable Palm Oil (RSPO). Unilever helped form the RSPO—an initiative working with the NGO, World Wildlife Fund (WWF), to develop global standards for producers which consumer goods manufacturers can purchase.

Realizing that having GreenPalm certificates was not doing enough, they took their commitments further; Unilever's goal is to purchase all palm oil from traceable growers to ensure that the palm oil used in their products is truly sustainably sourced. As of 2015, Unilever's palm oil was 73% traceable back to the actual mill that made the oil. The goal is to have 100% of all palm oil traceable and physically certified by 2020. This is quite an undertaking, actually having individuals on the ground in the region of source visiting and certifying the groves and mills to insure it is responsibly managed. In fact, they are now moving away from the GreenPalm certificates to this higher level of assurance.

Through a partnership with the World Resources Institute (WRI), a stringent "traceability and risk verification system on the ground with WRI's Global Forest Watch Platform" is being developed. This is an online monitoring and alert that uses satellite technology, open data, and crowd sourcing to guarantee access to timely and reliable information about forests.

Unilever is very transparent about their plans and pledges—on their website, they are making the following commitment: "From 2017, all of the palm oil we source will be traceable to known origins. Upstream suppliers would be required to provide independent third-party verification that the palm oil supplied to Unilever meets the principles outlined in their Sustainable Palm Oil Sourcing Policy if those sources are deemed high risk. From 2018, all of the palm oil sourced will be traceable to plantation. All non-smallholder sources of palm oil in the supply chain will be required to provide independent third-party verification to best practice standards that the oil supplied to Unilever meets the principles of our Sustainable Palm Oil Sourcing Policy. We will source all of our palm oil from physically certified or third-party verified sources" (Unilever 2017). These commitments put Unilever in a leadership position on this issue with some pretty big goals.

Water

Many of Unilever's products require the use of water, so a goal was set to halve the water associated with the consumer use of their products by 2020. As an example of how this is put in practice, the product Sunlight, a liquid dishwashing brand, is being used in water scarce areas. Research on dishwashing in India shows that consumers who use a liquid detergent instead of a bar use one-third less water, equivalent to saving two buckets of water every time dishes are cleaned. In addition to saving water, there is a social story here too, because the use of dish washing liquid has helped to free up women's time in developing nations and has changed their lives for the better.

Waste and Packaging

Purchasing over 2 million tons of packaging a year, a goal was set to halve the waste associated with the disposal of their products by 2020. Part of achieving this goal is the commitment to developing a circular economy through programs like the one in Indonesia. There is not a good infrastructure in Indonesia for recovering packaging materials, so, to address this, they established a Community Waste Bank Program, which empowers communities to collect and manage waste. This initiative has since been expanded into 19 cities and resulted in the collection of 3,739 tons of packaging waste that otherwise would not have been recovered.

Unilever's Greener Product Program

- Developed very aggressive long-term goals through their Sustainable Living Plan
- Strong CEO support for their program
- Significant commitment to sustainable sourcing
- Commitment to educating consumers to reduce their environmental impacts
- Partner with suppliers to drive sustainable innovations

Unilever has a large impact on the world's resources, purchasing significant amounts of raw materials and selling billions of consumer products. Having a company vision that embraces sustainability enables the development of greener products. The use of life-cycle thinking facilitates the focus on reducing environmental impact of the product at key life-cycle stages where it makes the most sense; in many cases, this is during consumer use or from raw material production. Unilever has developed education programs to help consumers reduce their environmental impacts. Being reliant on agricultural commodities, a very comprehensive sustainable sourcing program was initiated to partner with suppliers to reduce impact at the very beginning of their products' life cycle.

BASF

BASF is a large chemical company based out of Germany, with sales of 70.44 billion euros (2015), with 115,000 employees, and has production sites in about 80 countries worldwide. The main products sold are chemicals, performance chemicals, plastics, coatings, catalysts, crop technology, and oil & gas. Tag lines which have been used by the company, "We don't make a lot of the products you buy; we make a lot of the products you buy better®," and "We create chemistry" speaks to the nature of the company. As a chemical company, it's critical to focus on the safety and environmental impacts of your products. BASF believes this so much that they even embedded sustainability into their mission: "We create chemistry for a sustainable future." Even one of their market success principles is to "drive sustainable solutions."

When discussing their companies' key metrics, they report that 530 million tons of carbon emissions have been avoided through customer use of their products. BASF states that an objective of theirs is to "cooperate with our customers in creating and driving more sustainable solutions, which makes good business sense for both" (BASF 2015 Report). And that's one of the biggest drivers for having a greener product development program, delighting your customers by helping them with their sustainability goals!

Global Product Strategy

Being a member of the International Council of Chemical Associations (ICCA), BASF follows their Global Product Strategy (GPS). The GPS is a voluntary standard to ensure chemical manufacturers are following adequate product stewardship practices. The GPS requires very broad product stewardship initiatives:

- Establish a base set of hazard and exposure information adequate to conduct safety assessments for chemicals in commerce
- Perform a substance-specific risk assessment for all chemicals produced or sold worldwide
- Conduct training programs to build expertise in economies globally, with a focus on emerging economies
- Risk assessments are transparently shared with the public

Being transparent about their progress, BASF committed to making their risk assessments available to the public and to generate annual reports that detail their progress. The reports give an overview of the characteristics of chemicals as well as the different uses for the substance. The report includes hazardous properties of substances, safe handling methods, and the type and extent of potential exposure to humans and the environment (BASF 2017b).

Eco-Efficiency Analysis

BASF has been a pioneer in the concept of eco-efficiency which harmonizes the two concepts of economy and ecology. Eco-efficiency is an evaluation of the entire life cycle of a product or process "from cradle to grave, i.e., all the way from raw materials sourcing, to product manufacturing and use, to disposal." It also considers the consumption methods of the product and its end-of-life recycling or disposal activities.

BASF started its work on this process in 1996. They verified the validity of their approach through two independent third-party organizations, TÜV (German technical inspection and certification organization) and NSF

(National Sanitation Foundation). The purpose of eco-efficiency is to contrast different materials, environmental impacts across their life cycle.

The first step in their process is to evaluate environmental impacts in nine areas:

- Raw materials consumption
- Water consumption
- Land use
- Human toxicity potential
- Eutrophication
- Acidification
- Ozone depletion
- Photochemical ozone creation
- Climate change

Costs incurred in manufacturing and product use are included in the evaluation. Economic and ecological data are plotted on a graph, which is the eco-efficiency portfolio. The graph shows the comparison of a product or process compared with another product or process to show potential ecological and economical improvements or faults (BASF 2017a) (Figure 4.9).

For ecological evaluations, this analysis is based off of DIN EN ISO 14040 and 14044 which has been the standard since 2012. A third-party outsider,

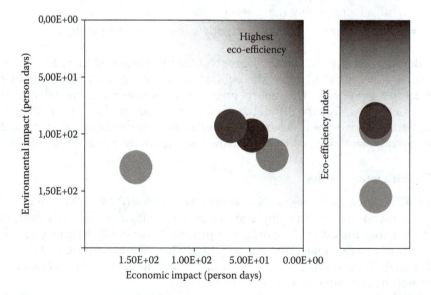

FIGURE 4.9
Eco-efficiency portfolio and eco-efficiency index (products in the upper right are the most eco-efficient).

the German Association for Technical Inspection, also validates the analysis and was most recently validated for the Eco-Efficiency Analysis in 2016, which marks 20 years that BASF's Eco-Efficiency Analysis has been used (BASF 2017a). To better understand how this evaluation assists in generating more environmentally friendly products, we can evaluate a few examples.

Eco-Package

There is a need for safer and more efficient packaging among farmers. BASF co-designed a lightweight plastic bottle for transporting pesticides with the farmers that used them. The collaboration resulted in a new design called the Eco Canister. The Eco-Efficiency Analysis showed that the new packaging has higher eco-performance than the original bottle. Pesticides are commonly used in farming, and now due to the new eco-packaging system, there is less risk for farmers to come into contact with the pesticide. Because it uses less plastic, it has lower environmental impact, resulting in less greenhouse gas emissions during the production and recycling processes. Greenhouse gas emissions have reduced about 2,000 metric tons of carbon dioxide by the use of the new Eco-Canister. The eco-efficiency chart shows that the new package is superior to the old package (Figure 4.10).

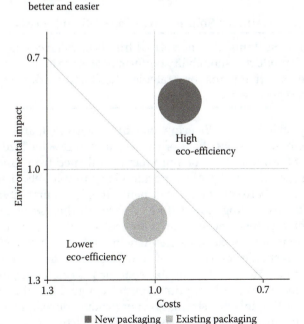

FIGURE 4.10
BASF eco-package eco-efficiency analysis.

Headline®

Headline® is a fungicide sold to enhance crop yields. An eco-efficiency study was completed for this product on corn sold in the state of Iowa (about 20% of all US corn comes from Iowa). The survey indicated that the use of Headline was a more eco-efficient solution, with a greater crop yield, lower environmental impacts, and lower production costs. When using this product, farmers will get the same amount of corn with fewer resources and less energy per bushel resulting in a 7% average yield increase and therefore a more eco-efficient product.

Sustainable Solution Steering®

BASF's greener product development platform is called Sustainable Solution Steering®. This process puts all products into four categories tied to providing sustainability solutions to their customers. The **"Accelerator"** category is the highest level and is described as products that provide a "solution with a substantial sustainability contribution in the value chain." **Performer**—meets basic sustainability standards on the market, **Transitioner**—has specific sustainability issues, which are being actively addressed, and **Challenged**—has identified significant sustainability concern and an action plan in development to address the concerns. As a chemical company, they believe chemistry is an enabler, offering "business opportunities" for meeting customer needs.

Sustainable Solution Steering® 3-Step Process

1. Analyze sustainability needs and trends of value chains
2. Check product sustainability performance in the market segments
3. Develop action plans for strategies, R&D and market approach; define concrete goal

---

The Sustainable Solution Steering® has three steps to it, and BASF has set a goal to increase the sales share of Accelerators to 28% by 2020. The way it works is that the value chain of a product is evaluated from cradle to grave considering industry and region-specific views in markets. They strive to achieve a balance between the three dimensions of sustainability: "Economy, e.g., potential cost savings for customers through the use of the products, Environment, e.g., ensuring standards are met, developing environmentally sound solutions, and Society, e.g., enhancing safety in production, use or end-of-life, stakeholder perception of solutions." The idea is to increase the portfolio of innovative and sustainable solutions (accelerators), making their customers more successful. Anytime you have a goal like that, it's a pretty good indicator that this is a strong greener product development approach. Adding on to this, the process has been assured by the consulting firm PricewaterhouseCoopers, which will give customers more confidence in this approach (BASF Sustainable Solution 2017).

The criterion for their highest level of classification "accelerator" is that each product must contribute to at least one of the following areas:

- Cost savings downstream
- Climate change
- Energy
- Resource efficiency
- Health and safety
- Biodiversity
- Renewables
- Emission reduction
- Water (BASF Sustainable Solutions 2017)

As an example of how Sustainable Solutions Steering works, let's look at how BASF addressed polyfluorinated substances. European authorities consider them to pose low risk to human beings and the environment, but BASF anticipated that based on trends, there will most likely be stronger regulation in the future. So polyfluorinated substances were rated as Challenged under the Sustainable Solution Steering® method when used in paper coatings. They decided to no longer sell these substances and, instead, developed paper coatings that do not accumulate in the environment and are biodegradable (ecovio®) or recyclable (Epotal®) that have been classified as Accelerators.

To give further insight into the type of products that are considered Accelerators, consider the category of products BASF sells to the automotive industry. Products like lightweight plastics which help cars achieve better fuel economy, fuel additives that improve the performance of cars, and catalysts which reduce pollutants are all Accelerator products.

BASF Product Stewardship Program

- Committed to product stewardship through participation in the voluntary Global Product Strategy
- Use their own developed Eco-efficiency tool to improve the sustainability of their products
- Have a greener product development approach called Sustainable Solution Steering® which classifies products in to four categories: Accelerator, Performer, Transitioner, and Challenged

---

BASF being a large chemical company has made assurances to their customers and stakeholders that their products will be both safe for human health and the environment. Commitments have been made to conform to voluntary international standards to insure products have adequate performance. To foster lower environmental impacts of their products, BASF has developed an eco-efficiency tool that measures the footprint of their products and cost benefits. This tool has been used to demonstrate the improved benefits of new products, helping to meet customers greener products desire. A greener product approach called Sustainable Solutions Steering is used to drive products into higher levels of performance, and goals have been set to increase their portfolio toward the highest sustainability rating.

H&M

The fashion apparel retail industry generates $25.3 billion dollars annually in the United States. This industry is one of largest in the world and is also one of the most polluting, second to the oil industry (EcoWatch 2017). One of the reasons why the fashion industry is so polluting is due to the extensive supply chain involved in making a garment. There are many steps necessary to bring a product to market and all of them have significant impacts: production, raw material, textile manufacture, clothing construction, shipping, retail, use, and ultimately disposal of the garment.

The carbon footprint in this industry is extremely large. In order to get one article of clothing to the consumer in a developed country, it typically travels thousands of miles since a lot of the manufacturing is performed far from the stores where the products are sold in. In addition to carbon emissions, process steps are quite polluting. Consider the pesticides used for cotton, plus wastewater, toxic dyes, and emissions from manufacturing synthetic textiles, and at the end of life of a garment, clothing that needs to be disposed of, just to name a few. The rise of fast fashion and over consumption has only increased sustainability issues for this industry. Apparel companies are starting to realize the effects of the fashion industry on the environment and are starting to take initiative to address these concerns.

H&M is a successful fashion retailing business headquartered in Sweden that generates $60.8 billion dollars (2016) in sales and is not only looking to become more sustainable but is focused on making the entire fashion industry more sustainable. H&M's reach affects the world because of the amount of stores (3,900), suppliers, and manufacturing plants they influence. The actions that they take impact many of the world's communities and the environment. In the apparel industry, there are many challenges to overcome when you consider the raw materials used for fabric and yarn production, and garment manufacturing. Perhaps the most impactful thing H&M can do is to choose lower impact fabrics and materials that minimize environmental concerns and to manage end-of-life issues regarding their products.

Sustainability Initiatives

H&M has a very extensive sustainability program, as evidenced by their 130-page sustainability report. Setting a high bar for other apparel companies, they have programs covering all aspects of sustainability including environmental objectives, supply chain, and ethical issues. As with other companies, setting company-wide goals is a critical part of propelling a company to develop greener products.

Goals

The company has set very extensive 2020 goals that cover product manufacturing, recovery, and employee efforts. Their plans are to make manufacturing greener through the use of 100% certified organic cotton or recycled cotton, eliminate the use of solvent-based glues in footwear and accessories production, use water-based polyurethane for shoe orders, use 100% man-made cellulosic fibers to decrease deforestation, and use certified sources for wool.

All H&M store locations are required to set up garment collecting services; they plan to set industry standards for the measurement of product sustainability, and engage employees by requiring sustainability training for H&M's colleagues, and lastly, engage employees and consumers in sustainability work. Operations are to focus on green power through renewable electricity and reduce their carbon emissions. In fact, in 2015 the company used an impressive 78% of all their energy globally from renewable electricity. They also decreased their total carbon dioxide emissions by 56% since 2012 (H&M 2017).

Embracing Circularity

H&M has established garment collecting services at all their stores, and they committed to water stewardship to act more responsibly with the earth's resources. Moving toward a circular model and closing the loop makes sure that the clothing has the longest possible use by recycling an old garment into a new one. They have become a leader in circular economy thinking by collecting about 22,000 tons of used clothing and reused about 1.3 million pieces of that clothing to make new garments (in 2015); they claim to be one of the biggest users of recycled polyester in the world (H&M 2017).

To encourage customers to recycle, H&M offers a £5 gift voucher to anyone who brings their unwanted clothes to one of their stores. Through this initiative, they hope to collect 25,000 tons of unwanted clothes per year by 2020. The funds from the clothes are donated to the H&M foundation, "with half going to research into textile recycling, and the rest to projects that focus on equality and marginalized groups" (Rains 2017).

Green Chemistry

The company is a large supplier of jeans, which requires massive amounts of water. H&M collaborated with Jeonologia consultancy experts to reduce their water use. In 2015, all of H&M's denim products reached "green" level using a maximum of 35 liters of water for every article of clothing.

Another issue that is of concern for apparel is dyeing fabrics, which requires the use of and generation of hazardous chemical waste and wastewater. H&M has committed to lead the apparel industry toward zero discharge of hazardous chemicals by 2020. The company also has a new chemical management strategy in which they are working toward controlling their chemical input and planning to phase out hazardous chemicals with healthier substitutes (H&M 2017).

H&M Greener Product Program

- Commitment to circularity by establishing collection and recycling of used clothing
- Have comprehensive sustainability programs addressing sourcing, manufacturing, and their products
- Use green chemistry thinking to improve production processes
- Have a greener product line called Conscious Clothing

Conscious Clothing

A noteworthy sustainability initiative is a greener product line called "Conscious" clothing that is made with about 50% organic or recycled materials, and/or other sustainable fabrics. On H&M's website, you can find a page called Conscious-Sustainable Style that is dedicated to providing the consumer with every-day sustainable clothing at an affordable price. The Conscious Exclusive collection gives the consumer the chance to purchase high-end pieces that are good enough to walk on the red carpet. H&M is trying to make a statement that high-end fashion can be sustainable too.

The fashion apparel industry has lots of room for improvement and is in need of changing their practices to become more sustainable. H&M has committed to leading their industry toward a brighter and healthier future. There are a number of goals and policies that have been put in place to ensure that the future of fashion becomes more environmentally friendly. They have a greener clothing line that uses recycled material and a commitment to a circular economy by collecting and recycling used clothing.

Common Practices among Leaders

By evaluating leading companies' activities, we can determine what the key initiatives are that they use to bring greener products to market. There are three main initiatives used:

1. Framework for product developers
2. Goals for developing greener products
3. Communication schemes

Framework for Product Developers

Leading companies believe that it is imperative to have methods that make it as straightforward as possible for product developers to take strides to make their products greener. Just about every company evaluated has a program to enable greener product design. The most frequent focus areas for these programs include reduction of energy, removal of toxic materials, reducing the product size and weight, packaging, and use of life-cycle thinking to identify the key aspects of a product to focus on environmental improvements. Also, the use of a scorecard is a leading practice that helps design teams to make the most meaningful improvements. Companies like Method, Philips, Seventh Generation, GE, and Johnson & Johnson all use focal areas and scorecards to help design teams to advance greener product design. In developing these frameworks, several companies have partnered with third parties to bring validity to their approach. Some have used consulting firms to assist in developing scorecards and others audits by third parties of their greener product approach.

Goals for Developing Greener Products

Just having a framework alone does not appear to be enough to bring greener products to the market. Most companies use sustainability goals focused on product improvements. Examples are revenue goals for greener products and R&D spending on sustainable innovations which GE, P&G, and Johnson & Johnson have all set. Several companies have goals that focus on sustainable sourcing, improvement in energy efficiency, removal of problematic materials like PVC and other hazardous chemicals, and management of the end-of-product life.

Communication Schemes

In order to inform customers of the positive attributes of their sustainable products, leading companies have developed ways to make it clear to customers that their products have improved performance. Branded green product lines like Ecomagination, Green Works, or Earthkeepers enable customers to quickly identify a greener product. Some companies have used innovative methods to communicate improvements, such as the nutrition-like label that Timberland uses. Others have looked outside their companies for assistance in communicating their greener products with third-party certifications such as the US EPA's Energy Star and Safer Choice certifications, or the Cradle to Cradle and FSC sustainable forest certification. Another communication method to gain the trust of customers is the transparency of materials and ingredients used.

Several companies have made bold commitments to transparency, agreeing to disclose every ingredient used. This is especially true for the deep green-based products like Method and Seventh Generation, but also with household product companies like SC Johnson and Clorox. Other innovative communication tools are product environmental reports that indicate various environmental and sustainability information. We will discuss these communication methods in more detail when evaluating the marketing programs companies use later in this book.

References

Apple Inc. 2010. A Greener Apple. http://www.apple.com/hotnews/agreenerapple/ (Accessed December 11, 2010).

Apple Inc. 2016. Environmental Responsibility Report 2016 Progress Report, Covering Fiscal Year 2015. https://www.google.com/search?q=Apple+Watch+Series+2+environmental+REPORT&oq=Apple+Watch+Series+2+environmental+REPORT&aqs=chrome.69i57.6949j0j8&sourceid=chrome&ie=UTF-8 (Accessed November 15, 2016).

Apple Watch Environmental Report. 2017. https://www.google.com/search?q=apple+watch+environmental+report&oq=apple+watch+environmental+&aqs=chrome.0.0j69i57j0.5979j0j8&sourceid=chrome&ie=UTF-8 (Accessed January 30, 2017).

BASF Annual Report 2015–2016. https://www.basf.com/documents/in/en/investor-relations/annual-reports/Annual%20Report%202015-2016.pdf (Accessed November 21, 2016).

BASF Booklet Sustainable Solution Steering. 2017. https://www.basf.com/en/company/sustainability/management-and-instruments/sustainable-solution-steering.html (Accessed February 5, 2017).

BASF Eco Analysis. 2017a. https://www.basf.com/en/company/sustainability/management-and-instruments/quantifying-sustainability/eco-efficiency-analysis.html (Accessed February 4, 2017).

BASF Global Product Stewardship Strategy. 2017b. https://www.basf.com/en/company/sustainability/management-and-instruments/responsible-care/product-stewardship-and-global-product-strategy/gps-implementation.html (Accessed February 4, 2017).

BASF Report 2015: Economic, Environmental and Social Performance. https://www.google.com/search?q=basf+report+2015&oq=basf+report+&aqs=chrome.0.0j6 9i57j0l4.4185j0j8&sourceid=chrome&ie=UTF-8 (Accessed February 5, 2017).

Ecomagination Strategy. 2016. https://www.ge.com/about-us/Ecomagination/strategy (Accessed October 13, 2016).

EcoWatch. 2017. Fast Fashion Is the Second Dirtiest Industry in the World, Next to Big Oil. http://www.ecowatch.com/fast-fashion-is-the-second-dirtiest-industry-in-the-world-next-to-big--1882083445.html (Accessed January 13, 2017).

GE Digital Industrial Annual Report. 2015. http://www.ge.com/ar2015/assets/pdf/GE_AR15.pdf (Accessed January 28, 2017).

GE Ecomagination. 2017. http://www.ge.com/about-us/ecomagination (Accessed January 28, 2017).

Get Your Software Kicks on Predix: GE Opens the World's First Industrial App Marketplace. 2016. http://www.gereports.com/get-software-kicks-predix-ge-opens-worlds-first-industrial-app-marketplace/ (Accessed September 21, 2016).

Greenpeace. 2010. About the Campaign. http://www.greenpeace.org/apple/about.html (Accessed November 25, 2010).

H&M Conscious Actions Sustainability Report 2015. 2017. http://sustainability.hm.com/content/dam/hm/about/documents/en/CSR/2015%20 Sustainability%20report/HM_SustainabilityReport_2015_final_FullReport_en.pdf (Accessed January 13 2017).

How Starbucks Saves Millions a Year in Energy with LED Lighting. 2016. https://www.greenbiz.com/blog/2010/12/02/how-starbucks-saves-millions-year-energy-led-lighting (Accessed September 27 2016).

Iannuzzi, Al. 2012. *Greener Products: The Making and Marketing of Sustainable Brands.* CRC Press. New York, p. 50,51, 59, 98.

Johnson & Johnson. 2015. 2015 Citizenship & Sustainability Report. p. 9. https://www.google.com/search?q=johnson+%26+johnson+sustainability+report+201 5&oq=johnson+%26+Johnson+sustainability+rep&aqs=chrome.0.0j69i57j0l4.27 326j0j8&sourceid=chrome&ie=UTF-8 (Accessed February 9, 2017).

Johnson & Johnson. 2017A. Earthwards® Strategic Framework. http://www.jnj.com/caring/citizenship-sustainability/strategic-framework/our-strategy-and-approach (Accessed February 7, 2017).

Johnson & Johnson Earthwards® Brochure. 2010. Brochure. Johnson & Johnson Inc. New Brunswick.

Johnson & Johnson Earthwards® Our Strategic Framework. 2017b. http://www.jnj.com/caring/citizenship-sustainability/strategic-framework/our-most-sustainable-products (Accessed February 8, 2017).

McDonough Braungart Design Chemistry (MBDC). 2016. http://mbdc.com/how-to-get-your-product-cradle-to-cradle-certified/ (Accessed November 17, 2016).

Method. 2016. Our Purpose, Ingredients & Packaging. http://methodhome.com/beyond-the-bottle (Accessed November 15, 2016).

New Product Standards Set Pace to Achieve 2020 Goals. 2016. https://www.timberland.com/responsibility/stories/new-product-standards-2020-goals.html (Accessed October 9, 2016).

OMNOVA Solutions' Green Bay. 2016. Wisconsin Paper Chemicals Facility Receives Environmental Award from General Electric. http://omnova.investorroom.com/index.php?s=43&item=389 (Accessed September 27, 2016).

P&G 2016 Citizenship Report. 2016. http://cdn.pgcom.pgsitecore.com/en-us/-/
    media/PGCOMUS/Documents/PDF/Sustainability_PDF/sustainability_
    reports/PG2016CitizenshipReport.pdf%22?la=en-US&v=1-201612122056
    (Accessed November 17, 2016).
P&G Brand Efforts. 2017. http://us.pg.com/sustainability/environmental-
    sustainability/brand-efforts. (Accessed February 1, 2017).
Philips Annual Report. 2015. https://www.annualreport.philips.com/downloads/
    pdf/en/PhilipsFullAnnualReport2015_English.pdf (Accessed October 30,
    2016).
Philips Green Products Sales. 2017. https://2015.annualreport.philips.com/#!/
    green-product-sales (Accessed January 29, 2017).
Rains, Holly. 2017. H&M's Bring It On Campaign is the Motivation You Need to
    Recycle Your Clothes. http://www.marieclaire.co.uk/news/fashion-news/
    hm-recycle-clothes-468996 (Accessed February 7, 2017).
Samsung Eco Bubble. 2017. http://www.samsung.com/uk/consumer/flagship/
    WF0806X8E/XEU/index_global.html (Accessed January 29, 2017).
Samsung Sustainability Management Framework. 2016a. http://www.samsung.com/
    us/aboutsamsung/sustainability/sustainablemanagement/sustainability-
    management-framework/ (Accessed November 12, 2016).
Samsung Sustainability Report 2016. 2016b. http://www.samsung.com/us/
    aboutsamsung/sustainability/sustainabilityreports/download/2016/2016-
    samsung-sustainability-report-eng.pdf (Accessed November 12, 2016).
SC Johnson. 2016. Focus on greener products. http://www.scjohnson.com/en/com-
    mitment/focus-on/greener-products/greenlist.aspx (Accessed October 27,
    2016).
SC Johnson 2015 Public Sustainability Report. 2016. http://www.scjohnson.com/
    Libraries/Download_Documents/SC_Johnson_2015_Public_Sustainability_
    Report.sflb.ashx (Accessed October 15, 2016).
Seventh Generation 2015 Corporate Consciousness Report. 2016. http://www.
    seventhgeneration.com/sites/default/files/2015_seventh_generation_
    corporate_responsibility_report.pdf (Accessed November 15, 2016).
Seventh Generation Multi Surface Cleaner. 2017. https://www.seventhgeneration.
    com/disinfectant-sanitizer?v=78 (Accessed January 30, 2017).
Sillydog. Mac mini, the world's most energy efficient desktop computer. http://silly-
    dog.org/forum/mac-mini-the-world-s-most-energy-efficient-desktop-comput-
    er-t15549s200.php (Accessed 2016).
The Clorox Company 2016 Integrated Annual Report. 2016. https://annualreport.
    thecloroxcompany.com/ (Accessed October 28 2016).
The Road to ecoROTR: How Building a Better Wind Turbine Began With an
    Online Shopping Spree for Styrofoam Balls. 2016. http://www.gereports.
    com/post/126500095500/the-road-to-ecorotr-how-building-a-better-wind/
    (Accessed September 21, 2016).
Iannuzzi, Al. 2012. *Greener Products: The Making and Marketing of Sustainable Brands.*
    Timberland's nutrition label via the Green Index®CRC Press. New York, p. 54.
Timberland. 2016. MEN'S TIMBERLAND® HERITAGE 6-INCH WATERPROOF
    BOOTS. https://www.timberland.com/shop/mens-6-inch-premium-water-
    proof-boots-wheat-10061024 (Accessed October 7, 2016).
Timberland Green Index. 2016. http://greenindex.timberland.com/about/ (Accessed
    October 7, 2016).

The Unilever Sustainable Living Plan. 2017. Greenhouse gases. https://www.unilever.com/sustainable-living/the-sustainable-living-plan/reducing-environmental-impact/greenhouse-gases/Our-greenhouse-gas-footprint/ (Accessed May 9, 2017).

The Unilever Sustainable Living Plan: Summary of Progress 2015. 2015. https://www.unilever.com/sustainable-living/the-sustainable-living-plan/our-strategy/ (Accessed February 4, 2017).

What this Turbine Does is Bigger than Winning the Triple Crown (or Prix de L'arc de Triomphe). 2016. http://www.gereports.com/post/120049112145/what-this-turbine-does-is-bigger-than-winning-the (Accessed September 22, 2016).

5

Advancement and Applications of the Framework for Greener Products

Alan Spray, Andrea Smerek, Chris Peterson, Doug Lockwood, James Fava, John Heckman, and Lauren Bromfield

A Framework for Greener Products Exists

Framework for Developing Greener Products (Fava 2012) sets the stage that tools, standards, and frameworks exist to embed sustainability considerations into the design, development, and commercialization of products. As has been described earlier in this book, the emerging issues and market expectations that companies face are increasing and expanding. The strategy implementation framework (Figure 5.1) presented in Fava (2012) allows for clear, conscious, deliberate definition of an environmental strategy with an understanding of the tools and systems that the strategy implies and necessitates providing a framework within which emerging issues can be managed.

Its foundations are a clear strategic intent and have been outlined (Figure 5.2) into four strategy options: *compliant, market driven, engaged,* and *shaping the future.*

These framework and strategy levels provide a flexible framework of systems, programs, and tools. Because the goals of various tools overlap, it is possible to link these approaches on a situation-specific basis through a flexible conceptual framework.

The evolution of environmental management strategies within businesses has been influenced by external drivers (e.g., regulations and customer requirements), internal drivers (e.g., cost savings and environmental strategies), and internal capacity and resources.

As firms move from a compliant strategy to a more sustainable strategy, different implications result. A *compliant* strategy, for example, is often viewed as a cost and often includes only strategic elements aimed at meeting the legal requirements as efficiently as possible. In a *market-driven* strategy, for example, a firm has integrated pollution prevention and customer/consumer or reactive market considerations into the design of its products or processes, which results in cost savings and/or cost avoidance. On the other end, a *shaping the future* strategy

FIGURE 5.1
Strategy implementation business framework. (From Fava, Framework for developing greener products, in Al Iannuzzi, *Greener Products: The Making and Marketing of Sustainable Brands,* 1st edn, CRC Press, Boca Raton, FL, 2012. With permission.)

may generate revenue by viewing the environment from a strategic perspective to identify new business opportunities and greener products. One advancement has been building alignment of the strategy levels and implementation of frameworks to various business values, that is, growing revenue, enhancing brands, reducing costs, and mitigating risks. This is expanded later in the chapter.

In Chapter 5 of *Greener Products* (Iannuzzi 2012), we concluded with five core walkaway messages that lay the foundation for a broader understanding of "the what and why" greener products are and will continue to be an increasing important component of a company's portfolio. As you will recall, the five core messages were:

Message 1: Sustainability issues are managed over the entire life cycle.

Frequently asked questions include:

1. What are the life-cycle stages associated with my product?
2. What materials are in my product and where do they come from?
3. What are the impacts at each of the life-cycle stages?

    This information can then be used to identify hot spots and areas to target for improvement opportunities. Our advancement to product-innovation tools section later in this chapter illustrates an example where the appliance sector developed product-sustainability standards to inform innovation teams on targets and to improve the next generation of appliances.

Message 2: Each product category and material has its own set of footprint and sustainability characteristics.

Three different value propositions based upon the unique characteristics of three materials (Five Winds International 2001) were defined.

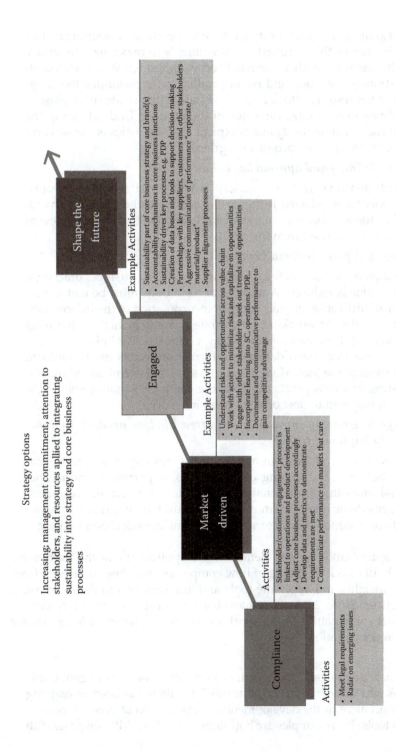

FIGURE 5.2

Strategy options for application within the strategy implementation framework. (Modified from Fava, Framework for developing greener products, in Al Iannuzzi, *Greener Products: The Making and Marketing of Sustainable Brands*, 1st edn, CRC Press, Boca Raton, FL, 2012.)

For example, a sustainability strategy for products containing metals (which retain their properties over time) is to maximize the utility of the metal through subsequent recovery and reuse. A sustainability strategy for wood and paper products is to maximize the integrity of the resource stock itself. A sustainability strategy for plastics might be to maximize the value of the product. Understanding the materials' sustainability and footprint characteristics is a basic early stage in any greener product program.

Message 3: Tools and approaches exist

As illustrated 5 years ago, a variety of tools exist and the appropriate tool can be selected to suit the questions and decisions at hand. Today, the toolbox has expanded and later on in this chapter a few of these are discussed in more detail.

Message 4: A great tool isn't enough

Without an owner or a defined place in an existing business practice, a tool's value is reduced. A number of conditions must be met to permit full utilization of tools and full incorporation of the information value in decision-making. The strategy implementation pyramid (Figure 5.1) illustrates that tools play a critical role within the framework—but tools need data and information that is used to inform and execute the tool. Moreover, unless the tool is embedded into an existing business practice within the company's management system, it may end up lost on someone's hard drive.

Message 5: Existing infrastructure often hinders product sustainability implementation

Implementing product sustainability measures requires communication between and among organizations, departments, and functional units that traditionally do not interact much. For continual progress to be made, we must move toward the integration of social and environmental concerns into core business decision-making.

This chapter builds on and demonstrates applications of the framework toward a better understanding of how companies have begun to learn from these key insights and apply the tools and approaches into their core business practices. Since 2012, there have been a number of advancements to the product sustainability framework as companies have made significant process implementation:

- *Product innovation tools*—As has been discussed throughout this book and *Greener Products* (Iannuzzi 2012), there has been an ongoing advancement in the development and application of product innovation tools. Two examples are highlighted. First, Reckitt Benckiser (RB) has developed a "Green Design tool" they call "Sustainable Innovation App" which can be considered a hybrid since it combines traditional quantitative life-cycle assessment (LCA) methods with qualitative "scorecarding." Often, a company has within their toolbox tools to ensure compliance with government regulations. Additionally, where there is market demand for broader sustainability, information (e.g., energy, greenhouse gases, water) from companies is a LCA tool (e.g., GaBi or SimaPro, Footprinter). The RB Green Design tool combined the product compliance issue with this broader life-cycle information into one tool. Second, the appliance sector developed and applied a "hot-spots analysis (HSA)" framework as a foundational step toward the development of a series of Product Sustainability standards that are being used to inform innovation teams on criteria which should be met in order for a product to be considered "greener."

- *Maturity models create framework to define "how" to improve*—With any emerging or existing sustainability issue, a number of questions, need to be asked:

1. *What* is the issue?
2. Is it relevant to my company, regions, and/or products? (the *why*)
3. *How* do I respond?

As can be seen through the review of social media, many companies have begun to understand the "what and the why" (although there is still much work to be done here). The challenge is "how do I respond." Each company may be at a different place along a continuum from ad hoc or just starting a sustainability journey to those that have already culturally embedded sustainability criteria into their business practices. These different levels of maturity are being recognized, and maturity models are being developed to inform a company of the various elements that should be included within their own strategy implementation framework and different levels of maturity.

- *Collaboration to advance performance and build capacity*—As a company moves along maturity levels, there can be topics or situations where working with other companies can influence the performance of each company but also the sector. The Electric Utility Industry Sustainability Supply Chain Alliance (Alliance) is one of these collaborations that has made excellent progress toward making their supply chain more sustainable.

- *Engaging the right resources to do the job: Managed services*—Any framework and toolbox are only as good as the people or resources that use or apply them. Companies are continually being asked to optimize resources against revenue and other business metrics.

Application of managed services or outsourced services to certain (but not all) sustainability practices has surfaced.

- *At the end it is all about change management—embedding a product sustainability tool—change management for success*—Many tools and data systems have been developed and remain unused or underutilized. Often problems develop because adjusting or changing an existing business practice would be required. The case study using Johnson & Johnson's successful implementation of its Earthwards® approach provides interesting insights and learning on change management success factors.

- *The ultimate success factor—linking product sustainability and business value*—Since publication of the earlier version of this chapter 5 years ago in *Greener Products* (Iannuzzi 2012), one of the most interesting and compelling advancements is an update of the simple graphic in Figure 5.2 (Fava 2012)—business implications on the strategy implemented. The ability to translate any sustainability plan, targets, and/or goals into the language of the decision-makers has been accelerated through the use of the business value framework. Does your greener product tool/plan contribute to growing revenues, enhancing your brand, reducing costs, or mitigating risks?

For each of these advancements to the original sustainability implementation framework, the following examples and insights are provided to inform the reader as to best practices and steps they can take to culturally embed sustainability.

Product Innovation Tools

Streamlined Product Innovation tool—The previous chapter described the various LCA approaches that exist today. Early adopters had focused on applying LCA within the development process, for example, Daimler and BMW. They provided excellent examples of how life-cycle information can be used to inform design. Within the product stewardship space, there is a continuum for ensuring the products are in compliance with governmental legal *requirements* to ensure a company's ability to sell their products within the countries they operate. Additionally, there *is a* growing demand for information on the environmental and social impact of products over their entire life cycle. Companies are looking for tools that address the continuum from product compliance to broader life-cycle information. RB has developed a "hybrid" tool that does just that.

Reckitt Benckiser's Sustainable Innovation App

Description of the Problem

Reckitt Benckiser (RB), one of the global health, hygiene, and home leaders, is all about performance. The company's betteRBusiness strategy has some big 2020 goals, including a third of their net revenue coming from more sustainable products and a third reduction in carbon and water impact per dose. To achieve this, RB recognized that everyone in the business had a role to play. They needed to tie these corporate-level targets back to product design.

RB's definition of "more sustainable" includes quantitative (LCA-like) assessment of carbon and water impacts and qualitative "attributes" such as an FSC-certified board. RB wanted to develop a tool that would be simple, easy, and accessible to everyone. A tool that combines qualitative and quantitative assessment gives real-time easy-to-understand results and delivers data/results in a manageable way that enables reporting and actionable insight.

RB began its journey in 2009 with a simple spreadsheet-based calculator that allowed innovators to understand the carbon impact of their design decisions. This helped RB achieve its 2020 target to reduce the carbon footprint of products by 20% per dose 8 years early. It also highlighted limitations—RB could not explore broader environmental impacts and evaluate trade-offs (e.g., carbon versus water), and it was heavy on administration. They ended up each quarter with several hundred spreadsheets in an email inbox and a huge task to pull it all together into reports, let alone generate insights for action.

The Solution

To scale RB's sustainable product development program to the demands of the business and incorporate additional metrics (e.g., water) that provide a broader view of the opportunities for innovation, RB worked with one of the authors and the software company Footprinter to develop a web-based solution to facilitate sustainable innovation. A screenshot from the RB App (as viewed internally) is illustrated in Figure 5.3. More information on RB's sustainable Innovation App can be found at RB (2017).

The Sustainable Innovation App is a web tool, allowing access to the latest version, by all R&D staff, wherever they are. It has a user-friendly interface that allows users to carry out assessment quickly and easily. Users build up their product by searching for raw and packaging materials and filling out details such as functional unit, material weight, recycled content, and consumer-use activity. Tags are applied to each assessment such as product category and brand. This allows for centralized roll-up and analysis across assessments. Users compare the impacts of their product against a "benchmark" product. Red/Amber/Green gauges show users

FIGURE 5.3
Sustainable Innovation App Screenshot, Footprinter.

Sustainable Innovation Calculator	Carbon g CO₂e / dose	Water Effective water L / dose	Packaging Effective packaging g / dose	Ingredients Self declaration
Better (More sustainable)	> 10% savings per dose	> 10% savings per dose	> 10% savings per dose	Complies with RSL + one new 'Preferred Sustainability Credential'
Same (No significant difference)	-1.5 – 10% savings	-1.5 – 10% savings	-1.5 – 10% savings	Complies with RSL
Worse (Less sustainable)	> 1.5% increase per dose	> 1.5% increase per dose	> 1.5% increase per dose	Does not comply with RSL (or variance)

FIGURE 5.4
Reckitt Benckiser Sustainable Innovation App scoring method.

how their product performs against the benchmark in real time. Overall, pass/no-pass simplifies results (Figure 5.4).

The tool is then embedded into RB's new product-design process, and designers have to assess innovations in order to progress through key Stage gates. This is critical as it has meant that sustainable thinking and design has been incorporated into "Business as Usual," and it is incorporated *early* in the design process. All the RB R&D teams are trained to use the tool; "champions" within each team are given extra training and carry out reviews/checks.

The engine that powers RB's Sustainable Innovation App is known as Footprinter's Green Design tool [https://footprinter.com/p/products/products/ (Accessed January 25, 2017)]. It is a framework which is customized for each company and can be launched in as little as two months. So while RB are particularly interested in mearsuring the carbon and water impact during consumer-use (a significant hotspot for their products), the tool has also been configured for a major tech hardware manufacturer to focus on the EU circularity metric. The tool has since been implemented for several large corporations. RB (and others) get their (non-sustainability specialist) R&D teams to use the tool. When being configured for non-specialist use the tool is kept simple with relatively little flexibility. Other companies employ a more complex version of Footprinter, this is used by internal sustainability teams as a customized LCA tool, and it has been verified to produce ISO 14040/14044 compliant LCAs.

The tool's flexibility means it has been used to implement some non-standard LCA metrics such as "naturalness" (% raw and packaging materials from natural sources) and the EU circularity metric. It also allows for combinations of quantitative and qualitative metrics, such as RB's "ingredients" metric where users qualify whether products have certain sustainability "credentials," such as FSC paperboard or Fairtrade certification.

Benefits—Sustainable and Business Value

RB's previous spreadsheet-based calculator was passed around its champion network to manually enter details of their innovations periodically for the sustainability team's review. This worked well, but it isolated insights with just a few of RB's team and wasn't always accessible early on in the design process.

RB's Sustainable Innovation App surpassed their previous solution in a number of ways:

- Enables designers to innovate and the sustainability team to engage and support, not just validate
- Intuitive centralized online tool with searchable product library
- Simple, visual, real-time traffic light indicators for each metric: carbon, water, packaging, and ingredients (sustainable attributes)
- Live comparison of results with other products/brand averages
- Ability to quickly modify existing product assessments, avoiding entering new data each time
- Hotspot-based focus for targeted innovations around the impacts that matter most
- View and report on status of brands and overall portfolio and share results across the business

Key Success Factors/Learnings

The Sustainable Innovation App was assured by the consulting firm EY and launched in November 2013. Within two months, more than half of RB's global research & development team were using it, with more than 700 product analyses completed. The tool is now fully embedded in RB's New Product Development (NPD) process. All new products have to be assessed through the app in order to progress through RB's NPD stage gates. At point of publication, the tool has 900 users with over 2,500 products analyzed. The results are used each year to report against RB's targets and it is also third-party assured. RB is on track to meet the target of a third of their net revenue from more sustainable products.

The success of individual product assessments has lead RB to expand the app's use into corporate level reporting. The tool is now used to complete LCAs for a basket of hundreds of "representative products" across RB's full portfolio. These representative products are then scaled up with sales data to estimate the total carbon and water impacts embodied in raw and packaging materials and from consumer use. This approach has been third-party assured and gives RB actionable insight into the sources of their impacts.

RB's goal was to have assessments take place in less than 30 minutes by non-LCA experts. RB has proven this is possible through the development of the Sustainable Innovation App.

Application of Hot-Spots Analysis to Inform Greener Products

The appliance sector has worked together to develop a series of product sustainability standards to drive improved performance. A key element of this effort was the development and use of HSA as a way to inform decision-making, specifically to identify those impacts beyond energy efficiency in the use stage based upon scientific information, results of LCA studies, professional and technical expertise, and stakeholder input.

Before we describe the appliance sectors experiences with developing and applying product sustainability standards and HSA, it would be useful to explain what is meant by HSA. UNEP (2014) described HSA as *"a methodological framework that allows for the rapid assimilation and analysis of a range of information sources, including life-cycle based and market information, scientific research, expert opinion and stakeholder concerns."* The flagship project 3a of the UNEP/SETAC Life Cycle Initiative titled *"Hot spots analysis—global guiding principles and practices"* has developed a methodological framework (UNEP 2017) to guide others on general considerations if one is interested in conducting a HSA. The methodological framework outlines eight steps to follow (Figure 5.5).

The project team received feedback and insights from stakeholders (UNEP 2014):

- Support "Beyond LCA" approach (qual. + quant. analysis)
- Clearly identify goal and scope
- Build stakeholder credibility and use a phased approach
- Keep hot-spots actionable and manageable
- Prioritize to meet goals for addressing hot-spots
- Make results clear and intuitive using visualization
- Periodically review and revise HSA
- Clearly communicate uncertainty
- Develop case studies and examples to support use

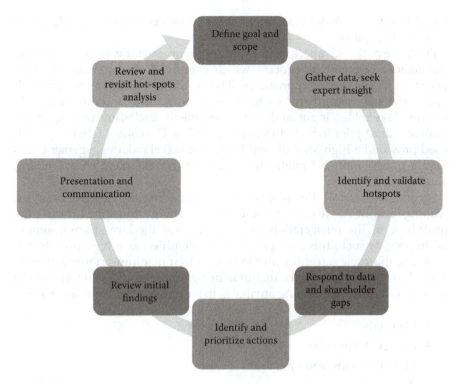

FIGURE 5.5
Methodological framework (From UNEP, UNEP/SETAC Life Cycle Initiative—Flagship Project 3a (Phase 2), *Hotspots Analysis: An Overarching Methodological Framework and guidance for product and sector level application*, Fava, J and Barthel, M, co-chairs, May 2017; Paris, France. http://www.lifecycleinitiative.org/new-hotspots-analysis-methodological-framework-and-guidance/.)

The appliance sector efforts to use HSA to inform its development of product sustainability standards, was one of over forty examples of HSA the project team encouraged during its initial Phase 1 mapping of existing HSA approaches. The appliance sectors efforts were interesting in that they developed a series of product sustainability standards for use by companies to inform each company's innovation practices.

As was realized through the results of the UNEP/SETAC Life Cycle Initiatives Flagship Project (FS) 3a project report, HSA goes beyond LCA. In other words, it can be based on more information than an LCA study, but can and often includes additional information from technical and professional experts. The appliance sector was able to utilize product life-cycle data, scientific studies, existing standards, industry and product experts, stakeholder concerns, and feedback from value chain players to develop a HSA that then informed the development of product sustainability standards (e.g., AHAM 7001-2012/CSA SPE-7001-12/UL. 7001 2012). Three methodological advancements stand out as valuable to the work of the sector, but also influenced the UNEP/SETAC hot-spots analysis project.

First, one of the advances used by the appliance sector was a collaborative stakeholder engagement process which combined characteristics of both private and consensus approaches. The resulting collaborative approach optimized the strengths of each approach to achieve a common goal to ensure stakeholder input and expert judgment, including testing of the draft standard prior to accreditation (Figure 5.6). The collaborative approach used provided a high level of credibility and stakeholder engagement, was highly transparent and relatively quick in its development of a standard, and kept costs down.

Second, the overall life-cycle screening (including the hot-spot analysis, scoping-level LCA, weighting) indicated that energy consumed during the use phase of the refrigeration appliance creates the largest environmental impact; as such, this unit process was identified as a high-priority hot spot. The life-cycle screening also identified four medium-priority hot spots related to the raw materials, manufacturing, and end-of-life phases of the appliance. The corresponding attributes to encompass these hot spots are:

- Materials
- Energy during use
- Manufacturing and operations
- Product performance
- End-of-life management

The most significant deviation from the LCA results is that the Energy Consumption during use attribute was allotted 45% of the weighting within this Standard by the authors and the appliance standards team,

Private	Consensus - based
Limited stakeholder input	*High-level of stakeholder input*
Low credibility	*High credibility*
Limited transparency	*Highly transparent*
Not typically accredited	Often accredited
Relatively short time to complete	Time intensive
Lower cost to develop	Higher cost to develop

Collaborative approach is a mix of two approaches

- Optimizes strengths of each approach to achieve common end goal
- Stakeholder input + expert judgment
- May involve testing of draft standard prior to accreditation

FIGURE 5.6
A collaborative approach.

compared to the greater than 70% share of life-cycle impacts according to the LCA. The team arrived at this value because it represents the largest share of any of the attributes (consistent with the LCA), while taking stakeholder input into account and encouraging manufacturers to make improvements in the areas covered by the other attributes.

Third, hot-spots analysis informed the prioritization of criteria that were used to develop specific targets which must be met in order for a new product to conform to the standard (e.g., AHAM 7001). But as was found to be relevant within the hot-spot analysis methodological framework, the hot-spot analysis informs on control or influence, that is, improve product performance in each criterion, thus resulting in a more sustainable product using the standard. This is exactly what happened, for example, Whirlpool was able to certify its KitchenAid 25 cu. Ft. 36"-width refrigerator to the AHAM 7001-2012 standard (UL 2012).

This example is an illustration of an advancement of the product sustainability toolbox to include development in the hot-spots analysis and its application within the appliance sector.

Collaboration to Advance Performance and Build Capacity

Sustainability by its very nature is complex and difficult to operationalize and implement. Multiple sectors and responsibilities are working to address this through collaboration platforms. A few examples include The Sustainable Apparel Coalition, Together for Sustainability, the Retail Industry Leadership Association, and the Sustainable Purchasing Leadership Coalition. By participating in a collaboration platform, individuals and organizations are able to scale resources around non-competitive efforts, share successful practice, and build on mutual lessons learned. In practice, participation in effective collaboration platforms leads to:

- An improved understanding of key issues and opportunities through both shared experiences and resources
- Accelerated performance from building on peer group models and lessons learned
- Development of tools and resources to codify successful practice
- Opportunities through benchmarking to identify risks and opportunities, as well as set priorities based on demonstrated value propositions
- Opportunities to pursue more ambitious and impactful change by working with other like-minded organizations

Two groups which demonstrate the practices and outcomes of an effective collaboration platform are the Electric Utility Industry Sustainable

Supply Chain Alliance (EUISSCA) or the Alliance and the Product Sustainability Round Table (PSRT). Both groups have an active membership group of approximately 15 member companies; they meet multiple times a year, operate within a pre-competitive space, have an explicit culture of collaboration, and are acutely focused on driving impact reductions through multiple channels. Below, we unpack the success and lessons learned from the Alliance as a model for individuals and organizations to seek out, develop, and leverage to accelerate their sustainability journey.

The Alliance, over the last 3 to 4 years has driven a tangible change within the industry, with members incorporating sustainability into decision-making, suppliers shifting from "check the box" levels of effort to innovative solutions, and a clear embracing of sustainability as a business priority throughout the supply chain. The Alliance's vision is *"to be known as the leader in establishing a robust and sustainable electric utility industry supply chain including advancing the maturity level of our members and stakeholders."* It is supported by the supply chain organizations of 15 leading electric utilities in the United States, including PG&E, DTE, Con Ed, Exelon, APS, Southern California Edison, etc. As a voluntary standards development organization, the development, implementation, and adoption of the standards are the anchor of all of the Alliance's activities. The success of the Alliance has been driven by multiple factors including committed members, a dedicated executive director and administrative support, a supportive supplier community, etc. However, five elements really stand out as being critical success factors. They are **three-year strategic plan, effective standards, a sustainability framework, an executive-level dashboard, and an engaged community of practice**. Information on these elements can be found on the EUISSCA website euissca.org (EUISSCA 2017).

The **three-year strategic plan** provides a common understanding of the direction, priorities, and goals of the group. This is complemented by yearly "roadmaps" that allow for adjustments to the pathway to the goals and are focused on execution. The **standards provide** a mechanism to document and codify successful practice on key material issues. A lot of effort is put toward ensuring the usability of the standards. The combination of these two factors create effective standards that stakeholders want, and are able, to apply quickly within their organizations.

The **sustainability framework**—and associated benchmarking against it—helps to illuminate the internal processes and practices needed to be successful in this space. Further, by applying a maturity-model approach, the logical steps between "initiating" and "transformational practice" are clarified. Finally, by having a strategic goal linked to progress against the framework, the members are motivated to improve and the Alliance, through the community, is empowered to support those efforts.

The **executive-level dashboard** has proven to be a critical key in unlocking the engagement of senior leadership, as well as focusing member activities on areas of need. It has provided a mechanism for Chief Procurement Officers to quickly understand their performance relative to their peers on a variety of initiatives within the Alliance. Further, by being able to inform members of the scope of activities, they are similarly able to identify and focus on priority areas quickly (Figure 5.7).

Finally, and fundamentally, critical to the success of the Alliance is the **engaged community of practice**. With members willing to share both their challenges and successful practices around the standards, the community is able to advance their practices, adoption of the standards, and realization of results more efficiently.

The key lessons learned over this time have been to build momentum, focus, on small wins that you have control over, build on those with increasing ambition, understand and address the obstacles to improved performance, and engage leading members on direction setting while supporting those who are not as advanced in order to leverage the resources of the community.

The Alliance provides an effective model that other industries, organizations, and practices should consider. Regardless of where one is on the sustainability journey, participation in a collaboration platform similar to the Alliance or PSRT can be a powerful catalyst to speeding and scaling one's activities and performance.

2018 goal progress

Goal status	Goal description
	Member maturity improvement
	Releasing three voluntary standards
	New practice adoption
	100% survey users embed results into supplier relationships
	Continually release materials to educate stakeholders
	Speak at four events per year
	Increase traffic to our website

On target	Off track with recovery plan	Off track, target will be missed

FIGURE 5.7
Sample EUISSCA 2018 2018 goal progress dashboard. (From EUISSCA, State of the Alliance 2016, Presentation, *EUISSCA Supply Chain Sustainability Conference*, New Orleans, LA, September 29, 2016. With permission.)

Maturity Models Create a Framework
to Define "How" to Improve

As emerging issues are identified and evaluated as to their relevancy to a business, the business needs to understand **what** they are and **why** they are relevant or not. Once they understand the **what** and the **why**, the next step is to determine **how** they can best build the infrastructure, tools, data systems, and capacity to incorporate that understanding and knowledge into their current business practices. Recognizing that businesses may be at different stages of their sustainability journey, and therefore different levels of maturity in understanding and embedding sustainability into business practices and activities, maturity models are being developed and applied.

A maturity model or "matrix" is designed to define maturity levels for specific management practices. The columns represent increasing performance, capability, organizational structure, and/or stability of programs and processes (e.g., from ad hoc, essential, or initiating to transforming or culturally embedded), and the rows represent specific management practices (e.g., prioritizing risks and opportunities, or product transparency and traceability) that an organization should have in place in order to fully embed sustainability into its organizational culture. Figure 5.8 below illustrates one example of a maturity-model framework for sustainability.

When determining which practices to pursue and how quickly, each organization must consider its individual circumstances and context, including its size, resources, strategy, priorities, and the estimated business value. Not all organizations will want to move across the model to the highest level of maturity for every practice. The key is for organizations to optimize their systems and programs based on their unique situation and goals.

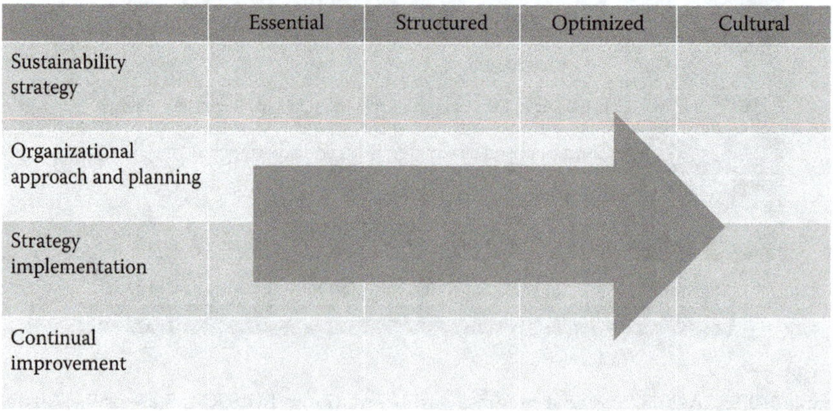

	Essential	Structured	Optimized	Cultural
Sustainability strategy				
Organizational approach and planning				
Strategy implementation				
Continual improvement				

FIGURE 5.8
Example maturity model from PSRT.

Once an organization decides where to focus its efforts, it can use a maturity model to inform how to improve within each focus area. For example, let's say an organization is starting at an "ad hoc" level in the area of supply chain management, and they have set a goal to progress toward the "transforming" level. While transforming is the ultimate goal, the organization will need to develop an action plan to get there. In some cases, the plan could involve drastic changes (e.g., business model change) that would allow the organization to "jump" directly to the transforming level. However, for the majority of companies, more gradual, step-by-step changes are required to progress through the various maturity levels.

Many organizations are adopting maturity model frameworks to embed sustainability into their organizational cultures.

Maturity models can be developed for various applications and can be broad (i.e., covering practices relevant to all organizations) or tailored to a specific sector or focus area (e.g., supply-chain management). The language used to define maturity levels differs and depends on the purpose and aim of the model. Although the titles of the levels vary, most maturity models use four or five levels of maturity.

Our case studies below illustrate how collaborative efforts among companies have developed and successfully applied maturity models to understand best practice, improve performance, and build more resilient organizations.

Electric Utility Industry Sustainable Supply-Chain Alliance (EUISSCA)

As highlighted in the collaboration section above, the Electric Utility Industry Sustainable Supply Chain Alliance (the Alliance) uses custom tools and systems, including their Environmental Sustainability Framework, to help improve the supply chain practices at each of the member utilities. The framework document provides an overview of the main elements or practices that are considered core to strong environmental sustainability management. The information is presented in a table that provides a means to gauge the level of maturity of corporate and supplier environmental sustainability programs, and to help shape continuous improvement goals for organizations seeking to advance their programs. The practices within the table, which are each defined by five maturity levels (moving from Initiating to Progressing to Optimizing to Leading and finally, to Transforming), were compiled by sustainability and supply chain subject matter experts from the Alliance member organizations and other professional organizations.

	Initiating	Progressing	Optimizing	Leading	Transforming
Scope of suppliers included in environmental assessment	Scope includes suppliers at an ad hoc basis.	Scope includes select top tier or strategic suppliers.	Scope includes majority of tier 1 suppliers (e.g., those represented by 80% or more of spend).	Scope includes majority of tier 1 and high impact tier 2 suppliers.	Scope includes majority of tier 1, high impact tier 2 and critical tier 3 suppliers.

FIGURE 5.9
One row within the Electricity Utility Industry Sustainable Supply Chain Alliance's *Environmental Sustainability Framework for Utilities.* (From EUISSCA, 2015, Electric Utility Industry Sustainable Supply Chain Alliance, Environmental Sustainability Framework, http://euissca.org/resources/utilities/environmental_sustainability_framework/. With permission.)

Figure 5.9 shows one row within the Alliance's framework (EUISSCA 2015), which defines the levels of maturity (columns) as follows:

- *Level 1: Initiating*—Organization has some awareness of sustainability with ad hoc activities.
- *Level 2: Progressing*—Organization has a systematic structure and formal processes.
- *Level 3: Optimizing*—Organization is advancing internally with continuous improvement.
- *Level 4: Leading*—Organization has embedded sustainability practices and is recognized for some leading industry practices.
- *Level 5: Transforming*—Organization has developed and implemented innovative practices that transform industry sustainability expectations.

Alliance members have benchmarked themselves against the framework (i.e., identified at which level they are performing for each practice) and have set goals to continuously improve in multiple areas. The framework has also helped to inform the development of voluntary standards for the Alliance that address gaps and improvement opportunities identified by members and encourage best-practice sharing between members.

Product Sustainability Roundtable (PSRT)

The Product Sustainability Roundtable, a global consortium of product sustainability leaders that engage in cross-value chain and cross-industry collaboration, has developed a maturity model or "Sustainability Leadership Framework" that can be used by any organization to embed sustainability into the organization. The Sustainability Leadership Framework has four levels of maturity (see Figure 5.8), including:

Level 1: Essential—Organization has basic elements needed to meet requirements and minimize risk. Responsibility is distributed to individuals, and a lack of programmatic processes means that practices are often ad hoc and inconsistently applied across the organization.

Level 2: Structured—Structured programs with common practices and procedures allow sustainability strategy to be implemented across the organization in a coordinated and consistent way. Responsibility is centralized into one or several dedicated experts, resulting in more efficient systems.

Level 3: Optimized—Systematic, organization-wide activities and tools are integrated into existing business processes and functions (e.g., product development and supply-chain management). Responsibility is shared among managers working together on implementation. Data analysis is used to optimize sustainability processes and performance.

Level 4: Culturally embedded—Sustainability is integrated into the business strategy and culture, and is considered in all business decisions and activities. Responsibility resides with top management and involves taking a life-cycle perspective and collaborating with value chain partners. Proactive methods are used to predict risks and opportunities and capitalize on business value (e.g., opportunities for innovation, competitive advantage, and brand enhancement).

As with other maturity models, the rows within the Sustainability Leadership Framework represent specific management practices (e.g., stakeholder engagement, planning for uncertainty, and value chain innovation). This framework guides users through four levels of maturity for each of the specific management practices that are highlighted.

The Sustainability Leadership Framework is used to accelerate and scale PSRT members' product sustainability efforts by:

1. Defining key management practices required to develop a leading, culturally embedded organization
2. Enabling self-assessment, identification of improvement opportunities, benchmarking against other member organizations and development of improvement goals
3. Identifying common challenges across the membership

4. Identifying best practices and resources that will enable continuous improvement within the framework

5. Providing a structure to measure performance and progress against goals over time

All organizations, regardless of where they are in their sustainability journey, can benefit from maturity models that define a path forward to building more resilient, agile organizations and ensuring continuous improvement over time.

Engaging the Right Resources to Do the Job— Managed Services

As companies look to downsize or "right size" as has it been sometimes called, a question that often surfaces is, How will we continue to meet the demands of the legal requirements and market demands. Do I build the capacity internally or buy externally? Obviously the answer is not easy. Perry, et al. (1993) outlined a simple decision logic that compared "proprietary and generic capabilities" with "is the work to be done value added or essential support work." For example, if the work is value-added support work but of a generic capacity, then the guidance was to develop ongoing access to the best capacity possible. However, if the work is value-added but proprietary in nature, then the guidance is to develop the best internal capacity possible. The authors have experienced this type of logic being applied within the product sustainability space. The following discussion provides insights into the rationale and lesson learned.

Ensuring compliance with material-related regulations, such as the EU Restriction of Hazardous Substances (RoHS) Directive and the EU REACH regulation, and driving toward greener products, requires a capable and efficient process to request and validate compliance data from the supply chain. The recognized best practice is to require suppliers to report full material disclosure (FMD) data (e.g., all substances present in each homogeneous material) for the components and materials they provide. Increasingly, companies are also requiring third-party laboratory analytical reports which quantify concentrations of analytes to verify compliance with EU RoHS and voluntary standards such as Low Halogen. Given the large number of components and materials that may be present in a product, this reporting process is highly transactional and potentially onerous from the standpoints of keeping up with the need for FMD data as new products are launched, validating that the data collected is complete and accurate and ensuring that the data collection process is conducted at the lowest cost of ownership.

Increasingly, this work is viewed as value-added but not proprietary in nature, and companies are looking at the opportunity to buy capacity externally as a managed service. This approach enables focusing corporate resources on strategic issues versus transactional work, at a lower cost than may be achievable via dedicated resources. Key considerations to ensure that the managed service meets corporate requirements for product compliance and development of greener products include:

- Standardization of supply-chain data requests around FMD data collection utilizing data exchange models such as IPC 1752A (Materials Declaration Management Standard. Latest revision: February 2014)
- Supplier on-boarding, training, and capacity building to ensure suppliers have the understanding and information needed to effectively respond to data requests
- Scalable workflows and supporting software to accommodate evolving requirements related to material disclosure, conflict minerals, and supply-chain sustainability data
- Integration with the customer's engineering systems and compliance system of record for compliance data. Alternatively, for customers who do not have an existing compliance system of record, establishment of a cloud-based system of record
- Leverage off-shore technical resources for day-to-day transactional work to reduce the cost of the managed service and to ensure resources are available in the time zones where suppliers are located
- Preparation of internal management reports as well as customer and regulatory reports
- Continuous improvement of business processes and software automation to flatten and reduce overall cost of ownership

A managed service should be set up under a Service Level Agreement (SLA) which articulates the services, technology, and key performance indicators which will be used to evaluate the managed service on an ongoing basis. An SLA is critical to ensuring that key parameters are understood and agreed to including that the desired outcomes of the managed service are achieved, which typically include cost reduction, improved supplier engagement, data quality improvement, support for metrics and reporting, compliance assurance, and reduction of product life-cycle impacts.

In the End, It Is All about Change Management—Embedding a Product Sustainability Tool—A Change Management Success

As the field of sustainability performance advances, decision-support tools allow more and more useful data to become available. However, even if one has useful tools, they are often not adopted by companies—not for an obvious reason, but often it is because *"it was not invented here or with me as a key stakeholder"* or *"I have done it this way for years and I do not want to change—why should I?"* and *"how will it allow me to get my job done faster and easier"* are often the responses given. Johnson & Johnson has developed its Earthwards® approach, an excellent tool/ system to inform innovation teams about hot-spots or priority areas for targets for improvement. Moreover, Johnson & Johnson has made it a requirement for all products going through the NPD process to use the Earthwards® approach. What were the actions, success factors, and approaches to overcome obstacles that Johnson & Johnson undertook to make Earthwards® an integral part of its business practice? These are examined in the case study below through a series of questions asked to the developers of the Earthwards® process.

Johnson & Johnson Example

What was done to embed Earthwards® into new product development processes?
First, we developed a product stewardship requirement to be a separate chapter in our EHS & Sustainability Standards. The Standards are business requirements which all operations must follow. Part of the requirements were that Earthwards® checks had to be integrated into the NPD processes of the various business units. The implementation of these checks varies based on the different processes to bring new products to market.

What were the obstacles and how did you overcome them?
The main issue with embedding sustainability into NPD is the concern with slowing down the development process. The key is to make the checks easy, fast, and meaningful. Getting the help of product stewardship professionals to give advice and insights to product development teams was beneficial. Web enabling our scorecard and simplifying the documentation required was important so that we didn't turn off brand teams that were applying for Earthwards® recognition.

What were the success factors?
There were several key success factors: for starters, getting buy-in from key R&D management to make meaningful checks, and embedding key questions into the NPD process was critical. Connecting Earthwards® to our sustainability goals was helpful to focus business units' efforts since we have a very robust reporting system for our goals (2020 goals are for 20% of JNJ revenue from Earthwards® products). Soliciting customer insight and demand for greener products via our standard voice of the customer business processes and sharing this with management helped bolster the business case. Doing internal and external marketing of our Earthwards® recognized products was critical—making people aware of new products that met our criteria that were being recognized helped to spread the word and create demand. Having our key customers aware of our approach and the things we are doing to meet their demands helped our management to see the benefit to their business and

their marketing programs, for example, highlighting Earthwards® products to customers during the request-for-proposal processes and in one-on-one meetings with purchasing managers.

If you were going to do this again, what would you do differently?
We learned a lot along the way: simplifying and web-enabling tools, tying Earthwards® to processes that were being used by business units, expanding our criteria to address business and customer needs, such as the removal of materials of concern and social issues all built momentum for Earthwards®. We would have liked to have made these enhancements faster and to have anticipated some of these things. Also, we now position Earthwards® as an innovation tool. Becoming more aware of the potential of sustainability as an innovation driver would have been a good thing to do earlier on in the process. If we had worked earlier with our procurement R&D organizations, it would have helped us to develop even more Earthwards® products.

The Ultimate Success Factor—Linking Product Sustainability and Business Value
Historically, there has been a perceived disconnect between improved product sustainability and additional value for business. The traditional rationale is that sustainability attributes aren't valued by the marketplace. However, there are also often quoted examples of very successful product sustainability programs that continue to grow and prosper more than 10 years after introduction. GE's highly successful Ecomagination program has generated "$160 billion in Ecomagination product revenue" and resulted in a "32% reduction in GE's GHG emissions from the 2004 baseline (GE 2017)." Johnson & Johnson's Earthwards program has recognized products that are valued at $9 billion as of 2015. Why do these programs continue and thrive if such a disconnect exists?

Part of the answer may lie in the very nature of product sustainability tools and practices.

There is a tendency within the sustainability space to focus on activities—number of LCAs, sustainability reporting, development of standards, scale of product reviews, etc.—rather than on **outcomes**, absolute impact reductions, displacement of less sustainable alternatives, etc. To drive the latter requires a commitment to driving activities through the additional phases of implementation, measurement, and improvement. A telling example is the use of an LCA. An LCA that is completed and put on a shelf does nothing to address value generation or impact reduction. By contrast, an LCA that initiates a process of engagement with product development teams, identification of opportunities and goals, measurement of improvements, and linkages to market demand is dramatically more valuable to an organization (Figure 5.10). In other words, many product sustainability efforts never get beyond assessment to the point of actual improvement—improvements that can matter to customers.

More to the point, the rise of sustainability awareness in the market has created substantial drivers in the consumer market, which in many instances

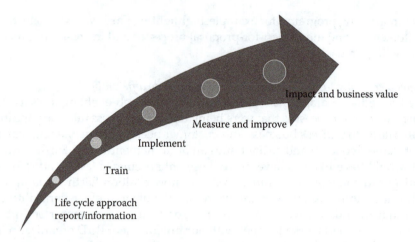

FIGURE 5.10
Outcomes to Impacts. (Adapted from Global Environment Facility (GEF). 2009. The ROtI Handbook: Towards Enhancing the Impacts of Environmental Projects - Methodological Paper # 2. www.gefeo.org)

may have not existed in the very recent past. The key realization to keep in mind is that every sustainability improvement does not resonate equally in every market. A recyclability claim may be very powerful for a packaging component that has very little perceived value to the end consumer who otherwise is faced with no other option than landfill. Conversely, recyclability for a longer lived energy using product may be of much lower importance than energy efficiency that contributes to lower cost to operate. Mapping received value related to sustainability onto specific market drivers can result in clear messages that resonate with customers.

In business-to-business markets, the realized value is usually easier to identify and explain. Most companies of all sizes are actively striving to improve their own sustainability performance, many of which are publicly reporting on their objectives and performance. These sustainability frameworks can be seen as a roadmap for suppliers to navigate, crafting sales messages to their customers of how their products can help them meet their goals. Some companies have built sophisticated programs to support this effort. A common theme among those programs is to combine resources to support both the R&D teams in reducing impacts, as well as the sales arm to effectively communicate the financial and sustainability benefits to customers. BASF provides an excellent example in their Sustainable Solution Steering program, which identifies and encourages development of "Accelerator" products that not only meet all sustainability requirements but also provide demonstrable benefit to customers. Known as *"Triple S,"* BASF quotes more than $16 billion of Accelerator solutions since the outset of the program, with more than 60% of their R&D spending being related to improving sustainability performance (BASF 2017).

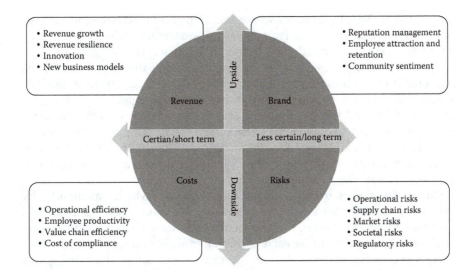

FIGURE 5.11
Business value framework. (Adapted from Esty, D. and Winston, A., *Green to Gold*, Yale University Press, New Haven, CT, 2006.)

To fully realize the potential for sustainability requires harnessing the systems, processes, and brainpower within one's organization. This in turn typically requires translating sustainability "lingo" and desired outcomes into the traditional business metrics of revenue, costs, brand, and risk. It is rare for any one activity to address all four elements equally; therefore, to truly optimize outcomes often requires focusing in on one or two areas as priorities. For example, a company that pursues a greener product strategy is likely to benefit from engaging their marketing team to help communicate the performance and impact reduction benefits from their products in order, internally and externally to realize results. The more that these contributions can be quantified and forecasted the better, but experience has shown that even operating within this framework can be powerful in aligning resources and realizing results (see Figure 5.11 for how the four elements fit together and should be considered).

Conclusion

During a recent conference in Japan, an exchange was held between a keynote speaker and someone from the audience; it went like this: The keynote speaker did an outstanding job in illustrating and demonstrating that businesses that perform well from a sustainability perspective also perform well from a business perspective. This was not a new message. It is conveyed almost daily in social media. Then, the speaker was asked, if the case is so strong for companies to create business value from operating sustainably, why isn't everyone doing it? Clearly a million-dollar question. Two answers surfaced.

Upon reflecting on that exchange, perhaps one of the reason is that the sustainability community need to learn to speak in the language of the decision-makers, as was articulated earlier in the chapter. Speaking to the CFO in terms of CO_2 reductions or gallons of water saved or a reduction in eutrophication does not resonate with someone who is thinking in dollars/euros/yen saved. Converting the CO_2 emissions reduction to one of the business values shown earlier grabs their attention and demonstrates the real value in adopting sustainable business practices.

During a workshop in New Zealand a few years ago, which was attended by over 40 business professionals, the business-value framework was described along with many examples. Each participant was asked before the workshop to think about a project that each of them had wanted to get funded which was unsuccessful. After a 4-hour interactive workshop, the participants were asked if they had previously known what they had just learned about the business-value framework and applied it to their project, whether they thought the project would have been funded—almost all said yes.

Second, meet them where they are. What does that mean? When the person who is facilitating a change steps into the shoes of the person who should be making the change, often changes are easier to achieve. The maturity model examples illustrate that by meeting them where they are, whether it be in product stewardship, supply chain, or a product innovation leader, starting where they are and moving toward a clear outcome with concrete mini steps makes the change more possible. For example, after a training of the Alliance's supplier maturity model, a sustainability vice president from a major supplier indicated that she loved the maturity model. She can now inform her operation resources that they do not have to go from the beginning to the final level of maturity in the dark—there is a clear path they can take, over time, to continuously improve and grow toward their goals.

In this chapter, we have advanced the strategy implementation framework through illustrations of how individual companies or collaborations of companies have used those advances within their own business practices to drive both sustainability and business value and performance forward.

References

AHAM 7001-2012/CSA SPE-7001-12/UL. 7001, The Sustainability Standard for Household Refrigeration Appliances.

BASF. 2017. Sustainable Solution Steering. https://www.basf.com/en/company/sustainability/management-and-instruments/sustainable-solution-steering.html (Accessed January 25, 2017).

Esty, Daniel and Andrew Winston. 2006. *Green to Gold*. Yale University Press, New Haven, CT.

EUISSCA. 2015. Electric Utility Industry Sustainable Supply Chain Alliance. Environmental Sustainability Framework. http://euissca.org/resources/utilities/environmental_sustainability_framework/ (Accessed December 10, 2016)

EUISSCA. 2016. State of the Alliance 2016. Presentation, EUISSCA Supply Chain Sustainability Conference. New Orleans, LA, September 29, 2016.

EUISSCA. 2017. *"About Us" Alliance Website*. www.euissca.org (Accessed January 25, 2017).

Fava, James. 2012. Framework for Developing Greener Products. In Al Iannuzzi, *Greener Products: The Making and Marketing of Sustainable Brands*. 1st Edition. CRC Press, Boca Raton, FL.

Five Winds International. 2001. Eco-Efficiency and Materials. International Council on Meals and the Environment, Ottawa, Canada.

GE. 2017. *A Decade of Ecomagination*. http://dsg.files.app.content.prod.s3.amazonaws.com/gesustainability/wp-content/uploads/2015/01/Ecomagination-Timeline-31.png (Accessed January 25, 2017).

Global Environment Facility (GEF). 2009. The ROtI Handbook: Towards Enhancing the Impacts of Environmental Projects - Methodological Paper # 2. www.gefeo.org

Iannuzzi, Al. 2012. *Greener Products: The Making and Marketing of Sustainable Brands*. 1st Edition. CRC Press, Boca Raton, FL.

Johnson & Johnson. 2017. Our Strategy and Approach. www.jnj.com/caring/citizenship-sustainability/strategic-framework/our-strategy-and-approach (Accessed January 25, 2017).

Perry, Stott, Responsibility Products. Smallwood. 1993.

RB. 2017. http://www.rb.com/responsibility & https://www.rb.com/responsibility/products/ (Accessed January 25, 2017).

UL. 2012. http://www.ul.com/global/documents/offerings/businesses/environment/press/ULE_whirlpoolcertifies.pdf (June 14, 2012) (Accessed January 25, 2017).

UNEP. 2014. *Hotspots Analysis: An Overarching Methodological Framework and guidance for product and sector level application*, Fava, J and Barthel, M, co-chairs, May 2017; Paris, France. http://www.lifecycleinitiative.org/new-hotspots-analysis-methodological-framework-and-guidance/.

UNEP. 2017. UNEP/SETAC Life Cycle Initiative—Flagship Project 3a (Phase 2). Hotspots Analysis: An Overarching Methodological Framework. Fava, J and M. Barthel, co-chairs. Copyright © UNEP DTIE (2014). Draft Final—to be finalized by February 2017.

6

Valuing Natural Capital

Libby Bernick*

Introduction

Companies have an enormous opportunity ahead: Consumer demand for products is set to grow significantly over the next 20 years with the increase in middle class consumers, especially those in emerging markets, who will require food, clothing, housing, and transport at levels never seen before. At the same time, this presents an enormous challenge for businesses to provide these products and services against a backdrop of increasingly scarce resources, such as water. Successful business models will be those that understand the value of the natural systems that provide these resources— commonly referred to as natural capital—and how these systems can be managed as part of the production of greener products and services.

Nature is an important part of the global economic engine and the brands behind it. For example, the US forest products industry is 4% of the total US manufacturing GDP, makes over $200 billion in products annually, and provides about 900,000 jobs (American Forest and Paper Association 2016). Nature also provides services to businesses, such as the clean water that a pulp and paper-making mill might require, which are much harder to value.

But in 2013, executives at the UN-backed organization, The Economics and Ecosystems of Biodiversity, set out to quantify just how much the economy relies on goods and services from nature. The landmark study, *Natural Capital at Risk—the Top 100 Externalities of Business*, found that primary production and processing industries (agriculture, forestry, fisheries, mining, oil and gas exploration, utilities; primary processing industries: cement, steel, pulp and paper, and petrochemicals) cost the economy around $7.3 trillion a year in terms of the economic costs of environmental impacts, things such as greenhouse gas emissions, loss of natural resources, loss of nature-based services such as carbon storage by forests, or air pollution-related health costs (Natural Capital at Risk 2013). At a cost of about 10% of the $75 trillion global economy, nature's contribution is vital to future business growth and viability.

---

* Contributions from James Richens and Sarah Aird

The analysis was the first to quantify and value "hot spots"—areas of greatest environmental impact for the economy, broken out by region and commodity. Figure 6.1 illustrates the top five commodity regions, the environmental costs associated with production, and the revenue generated by the business activity. The highest environmental costs are associated with energy generation and food production. These costs include the damage to society from impacts of greenhouse gas emissions, the health costs and societal damages from air pollution (for instance, asthma or crop damage), or water pollution from the excessive use of fertilizer which creates a cost for businesses or communities who have to pay to treat the water so they can safely use it. Topping the rankings are coal-fired power in Eastern Asia (ranked first with $453 billion in environmental costs) and in Northern America (ranked third with $318 billion in environmental costs). Another high-impact sector is agriculture, especially in water-scarce areas and where production and land use is also high. For example, cattle ranching in South America at an estimated $354 billion, ranks second while wheat and rice production in Southern Asia rank fourth and fifth, respectively. Even though many of these natural capital costs occur in the developing world, the resulting products are consumed by supply chains in developed economies, making it a global challenge for a globalized world.

Although the total costs in the Natural Capital at Risk analysis were strikingly high, perhaps the more surprising revelation is that the natural capital

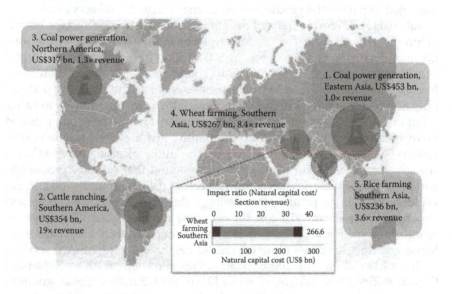

FIGURE 6.1
Natural capital costs of top impacting sectors in the global economy. (Courtesy of Trucost, London, UK.)

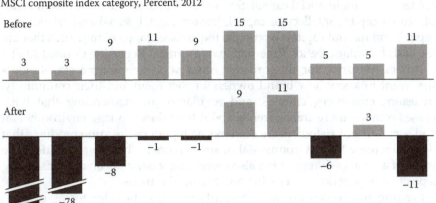

Profit margin (EBIT) before and after natural capital costs, based on top-2 companies in MSCI composite index category, Percent, 2012

FIGURE 6.2
Profit margins before and after accounting for natural capital costs. (From Stuchtey, M.R., Enkvist, P.-A., & Zumwinkel, K., *A Good Disruption, Redefining Growth for the Twenty First Century*, Bloomsbury, London, 2016.)

cost is higher than the total revenue of each sector. In other words, **not one of the top 100 businesses would be profitable if they had to pay the environmental costs associated with production**. Figure 6.2 compares the profit margin before and after considering these natural capital costs, vividly illustrating how business profitability depends on the flow of natural capital assets (Stuchtey et al. 2016).

The scale and variation in the natural capital costs across regions and commodities in the Natural Capital at Risk analysis show that companies have a very real opportunity to differentiate their brands, business models, and greener product manufacturing processes by incorporating natural capital valuation to optimize their supply chains, sourcing decisions, and product designs.

Creating Business Value

Historically, accounting systems have relied solely on traditional measures of financial wealth, for example, GDP as an economic measure or a profit and loss statement for a business. Increasingly, business leaders are recognizing that the value they create should be measured by a wider set of measures than shareholder return on financial capital. Six types of capital—financial capital, manufactured capital, intellectual capital, human capital, social and relationship capital, and natural capital—comprise the broader range of measures that are the basis for value creation (International Integrated Reporting Council 2016).

Part of the drive for this broader framework comes from stakeholders that want to know how brand owners impact them and their community. Investors, customers, citizens, and regulators are demanding that businesses provide more transparency about how they manage environmental and social risks. Customers are demanding higher performing products that do not increase the environmental or social costs to their communities. The value of a business's reputation also increasingly depends on demonstrating a positive impact on society—beyond financial returns.

Leading businesses are now quantifying these broader forms of capital and value creation. As an example, driven by a desire to understand the full value it delivers to society, the large Australian water utility, Yarra Valley Water, published the first-ever integrated profit and loss report quantifying its positive and negative natural, social, human, and financial impacts in monetary terms (see Figure 6.3). Its approach was to consider

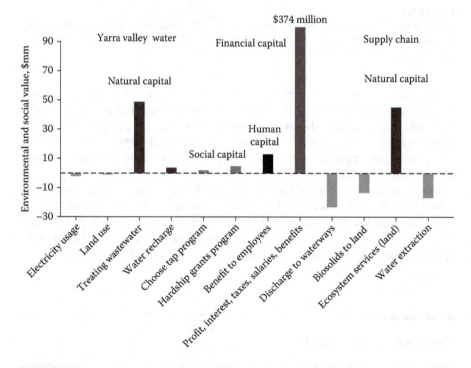

FIGURE 6.3
Integrated valuation of social, natural, and financial capital. (From Trucost Analysis using Yarra Valley Water data [2016].)

"value addition" through a holistic lens: Measuring impacts across all capitals and across all categories of stakeholders affected—including employees, customers and society at large—and not just its shareholder, the Victorian Government. Value is degraded when pollutants are discharged into the environment, waste is sent to landfill, and natural land is converted. Value is created when society receives free benefits, such as when water is cleaned and recharged back to the environment, when ecosystem services are provided, or when vulnerable families receive waivers on water utility bill payments (Yarra Valley 2016).

Natural Capital and Business Value

Natural capital, one of the six forms of capital, is the stock of renewable and non-renewable resources (such as plants, animals, air, water, soils, minerals) that provide a flow of benefits to people and the economy. Figure 6.4 illustrates the relationship between these natural capital assets and the resource flow to the economy, and how different activities (e.g. energy generation

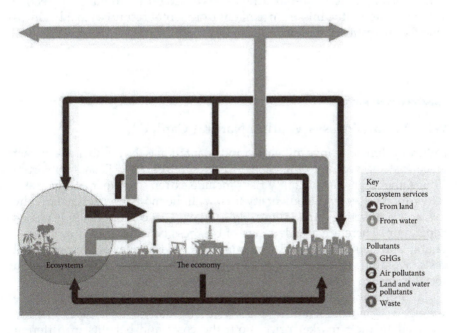

FIGURE 6.4
Economic and natural capital relationship. (Courtesy of Trucost, London, UK.)

or transportation) that use the resources then generate pollutants and emissions.

Some of the economic benefits provided by natural capital are valued, for instance, commodity raw materials such as wood. Other goods and services that nature provides are equally important to the economy yet not fully valued by markets; for example, clean water that a business needs to run its production processes, or the clean beach that an airline might depend on to book seats to resort destinations. The term natural capital cost is used here to describe the non-market value of the environmental resources that businesses depend on to grow revenue.

Because these natural capital-related goods and services are not fully valued by markets and are not part of corporate finance accounting systems, they are typically not considered in most business decisions. Imagine a group of high-use commodities that drive enormous value in the economy, but where there are no price-setting participants because there is no market. Then think about what the potential is for mispricing and dramatic price volatility of these resources and the implications for business performance. In the most extreme case, the resources disappear before the price can move because there is no price signal to curtail demand. That is the precise situation for the vast majority of the world's essential environmental resources, from clean air and fresh water to natural habitats, landscapes, and a stable climate. The current situation results in inconsistent or incomplete decision-making that is based on anecdotal or subjective judgments about the importance of environmental resources to a business model or investment decision.

Why Are Businesses Valuing Natural Capital?

Failure of climate change mitigation and adaptation is the number one impact global risk, according to the Global Risks Report (World Economic Forum 2016). Other high sustainability risks include extreme weather events, water crises, food crises, and biodiversity loss, with the most severe risks faced by the most productive—and vulnerable—regions that company supply chains depend on. There are a wide range of business benefits to valuing natural capital, outlined in Figure 6.5, which include, for example, lower operational costs, avoiding regulatory risks, reducing lending costs, or maintaining a social license to operate.

Making, using, and disposing of products—even greener ones—have a range of environmental and social costs that in most cases are not reflected in their market price. Both the costs and benefits of different material options need to be considered to enhance the sustainability of

Operational	• Reduce raw material costs and interruption to supply chains from weather, flooding, etc.
Regular business activities, expenditures, and performance	• Realize efficiency gains
Legal and regulatory	• Identify future legislation
Laws, public policies, and regulations that affect business performance	• Reduce compliance costs and risks of fines and penalties
Financing	• Reduce financing costs and increased margins
Cost of and access to capital including debt and equity	• Improve access to financing–attracting investors
Reputational and marketing	• Identify new revenue streams and differentiate your products
Trust and relationships with stakeholders, customers, employees, and suppliers	• Improve ability to attract and retain employees
Societal	• Identify benefits and negative impacts to local communities through improved natural capital (e.g., water quality)
Relationships with wider society	• Support a social license to operate

FIGURE 6.5
The business benefits of valuing natural capital (Natural Capital Coalition).

these products. Applying environmental or "natural capital" valuations allows brand managers and product designers to measure and communicate environmental impacts in monetary terms.

A Proxy for Risk

Externality costs are an excellent proxy for business risk. Natural capital costs affect the future profitability of businesses if they are internalized as a business cost because of increased regulation (e.g., a tax on carbon emissions or reduced water allocations) or because of pressure from customers and communities concerned with the impact of products. These natural capital costs are especially important for industry sectors with small profit margins.

Quantify and Understand Environmental Impacts in Business Terms

Putting a monetary value on impacts translates environmental issues into business terms and allows businesses to consider their environmental costs alongside financial costs. Showing all impacts enables businesses to identify hot-spots, uncover unintended impacts, and compare impacts between internal business units or even across the different forms of capitals.

Evaluate Trade-Offs

There are different types of environmental impacts (e.g., air pollution or water pollution) and each is expressed in a different unit, from CO_2 equivalents to gallons of water used. For example, if greener brand A has higher greenhouse gas emissions but lower waste compared to greener brand B, which is better? Natural capital valuation places a monetary value on each impact, so they are all expressed in financial values. In this way, impacts can be added and compared, and trade-offs between different design strategies become clear.

Science-Based, Context-Based Insights

Natural capital valuation is also a good way to communicate about more sustainable brands in an objective way. Because natural capital valuation accounts for the availability of the resource, business can understand the context for its resource use. For example, a company that uses water from a stressed aquifer has a much greater impact than if water is used from a plentiful region. In addition to providing science-based environmental context, environmental costs are presented in a business context that allows companies to understand how material costs and risks impact future profitability.

Communicate in Business Terms for Investors, Customers, and Consumers

Communicating about the benefits and impacts of sustainable brands in monetary terms helps engage customers and consumers who are not environmental experts. For many people, nature can be a bit abstract. Seeing the true cost of a product alongside the financial cost shows that these environmental impacts are real, they are being paid for by society, and they drag down our economies. As stakeholders expect more transparency about company business practices, natural capital valuation is also a good way to report on corporate sustainability policy and progress in a more objective manner that is also aligned with the Global Reporting Initiative, Sustainability Accounting Standards Board, and the International Integrated Reporting Council.

Natural Capital Valuation Standards, Protocols, and Frameworks

There are many approaches to measure and value impact as shown in the following table, which illustrates how a wide range of organizations are applying natural capital accounting at different scales and for different purposes. At the same time, public and private sectors are working to harmonize and align approaches for natural capital accounting.

Examples of Natural Capital Valuation

Scale or Type of Application	Example Initiatives
Sovereign	UN System of Environmental Economic Accounting (SEEA). Wealth Accounting and Valuation of Ecosystem Services (WAVES), a global partnership led by the World Bank to ensure that natural resources are accounted for in development planning and national economic accounts (Wealth Accounting Valuation Systems 2016)
Financial Lending	In 2015, TD Bank valued the natural capital impact related to reduction of greenhouse gas (GHG) emissions and airborne pollutants of TD Bank's $200 million green bond (TD Bank 2015)
Commodity	American Chemistry Council in 2016 valued the costs and benefits of plastic consumer products and packaging (American Chemistry Council 2016)
Corporate Enterprise	Kering (the French luxury goods holding company which owns Alexander McQueen, Balenciaga, Gucci, Puma, Volcom, and other sport and lifestyle brands) developed the concept for an Environmental Profit and Loss framework in 2011 and has since applied it across the operations and supply chain of its enterprise (Puma 2011; Kering 2015)
Product	Dell in 2015 valued the net benefits of its closed-loop plastic recovery system (Dell 2015)
Project, Asset or Site Level	In 2016, Dow committed to quantify and implement business-driven project alternatives that will best enhance nature and deliver $1 billion in net present value (Dow 2016)

National level initiatives include the UN System of Environmental Economic Accounting implemented by governments through, for example, the World Bank-led Wealth Accounting and Valuation of Ecosystem Services global partnership. The Natural Capital Coalition and Natural Capital Declaration focus on approaches aimed at business and investor decision-making.

The Natural Capital Coalition is a global multi-stakeholder collaboration that brings together leading global initiatives and organizations to harmonize approaches to natural capital. The Natural Capital Protocol, launched in 2016, provides a framework designed to help generate trusted, credible, and actionable information for business managers to inform decisions. The Protocol is broadly applicable across all businesses and also includes sector-specific guides for the apparel industry, food and beverage industry, and financial services industry (Natural Capital Coalition 2016).

Because there is a growing trend in monetary assessments by government (e.g., regulatory cost benefit analysis or the polluter pays principle) and private industry (e.g., in reporting or risk assessments) the International Organization for Standardization (ISO) has announced plans to develop ISO 14008, a framework for the monetary valuation of environmental impacts. The proposed standard will provide a common framework and terms and increase the transparency behind the natural capital costs and how they are determined.

Natural Capital Valuation Frameworks

Corporations apply natural capital accounting using a wide variety of frameworks depending on the business decision being informed, the scope of the application, and how it will be communicated.

One of the first frameworks developed was the groundbreaking Environmental Profit and Loss framework created by Puma in 2011 (see Figure 6.6), intended to mirror a conventional Profit and Loss statement by quantifying environmental costs in monetary terms. For Puma, environmental costs totaled €145 million and were largely concentrated in cattle-rearing and cotton-farming activities in the upper tiers of its

EUR million	Water use	GHGs	Land use	Other air pollution	Waste	Total	% of total
	33%	33%	25%	7%	2%	100%	
Total	47	47	37	11	3	145	100
PUMA operations	<1	7	<1	1	<1	8	6
Tier 1	1	9	<1	1	2	13	9
Tier 2	4	7	<1	2	1	14	9
Tier 3	17	7	<1	3	<1	27	19
Tier 4	25	17	37	4	<1	83	57

FIGURE 6.6
Puma Environmental Profit and Loss (EP&L) framework. (Courtesy of Trucost, London, UK.)

supply chain (Puma 2011). This type of framework is especially well-suited to understanding corporate enterprise-wide risks, because it aggregates natural capital costs across all operations and supply chains and can be broken out by environmental impact, business unit, region, or commodity.

A net benefit framework is also commonly used to apply natural capital valuations and is especially suited to quantifying the reduced environmental costs of greener product design or sourcing. A net benefit framework may include a net present value calculation for long lived, capital-intensive projects, or can also be extended to show the overall economic and societal benefits of a product or system that is applied at scale within a market segment or supply chain.

Valuation of Natural Capital Costs

Regardless of the framework that is applied to organize or communicate the insights, the natural capital valuation methodology generally follows three steps (Figure 6.7):

- Quantifying emissions or resources used (e.g., metric tons of air pollutants emitted or gallons of water used),
- Quantifying the positive or negative (e.g., decreased biodiversity due to water scarcity or decreased life expectancy due to air pollution), and
- Valuing in monetary terms the costs of the impact to society.

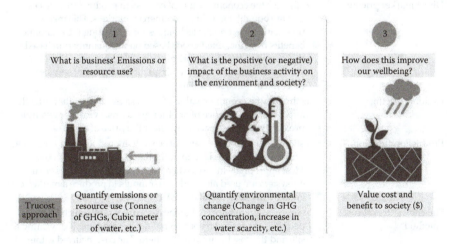

FIGURE 6.7
Approach to valuing natural capital impacts and dependencies. (Courtesy of Trucost, London, UK.)

Businesses typically use tools such as carbon or water footprints, life-cycle analysis, impact pathway analysis, or material flow analysis to quantify the amount of resources used and their impact on society, the environment, or the business itself. Each impact can have several consequences. For example, water depletion can affect society (loss of drinking water or decreased food supply), as well as the environment (not enough water to sustain fish and wildlife) and business itself (lack of clean fresh water that requires the business to install treatment facilities). For each impact, Trucost develops a valuation coefficient that reflects the cost or benefit and associated use of inputs and emissions on natural and social capital. A variety of standard economic techniques, as shown in the table below, are used to value natural capital costs depending on the data available and specific application or business decision.

Overview of Valuation Methodologies

Valuation technique	Description
Abatement cost	The cost of removing a negative by-product, for example, by reducing the emissions or limiting their impacts.
Avoided cost/ replacement cost/ substitute cost	Estimates the economic value of ecosystem services based on either the costs of avoiding damages due to lost services, the cost of replacing ecosystem services, or the cost of providing substitute services. Most appropriate in cases where damage avoidance or replacement expenditures have or will be made (Ecosystem Valuation 2000)
Contingent valuation	A survey-based technique for valuing non-market resources. This is a stated preference/willingness-to-pay model in that the survey determines how much people will pay to maintain an environmental feature.
Direct market pricing	Estimates the economic value of ecosystem products or services that are bought and sold in commercial markets. This method uses standard economic techniques for measuring the economic benefits from marketed goods based on the quantity purchased and supplied at different prices. This technique can be used to value changes in the quantity or quality of a good or service (Ecosystem Valuation 2000)
Hedonic pricing	Estimates the economic value of ecosystem services that directly affect the market price of another good or service. For example, proximity to open space may affect the price of a house.
Production function	Estimates the economic value of ecosystem products or services that contribute to the production of commercially marketed goods. Most appropriate in cases where the products or services of an ecosystem are used alongside other inputs to produce a marketed good (Ecosystem Valuation 2000)
Site choice/travel cost method	A revealed preference/willingness-to-pay model which assumes people make trade-offs between the expected benefit of visiting a site and the cost incurred to get there. The cost incurred is the person's willingness to pay to access a site. Often used to calculate the recreational value of a site.

Many environmental and social impacts are site specific because they depend on local conditions. The ideal situation is to use site-specific primary data, however, many times this is not possible given budgets, schedules, or technical constraints. As an alternative, scientists may use the value transfer method. In this method, the economic value is estimated by transferring available information from completed studies to another location or context by adjusting for known variables (such as population density, income, or ecosystem type and size).

Best practice guidelines for value transfers have been set out by UNEP's Guidance Manual on Value Transfer Methods for Ecosystem Services (Brander 2013).

Case Study Examples

In 2016, over 750 organizations publicly disclosed their involvement in natural capital valuation initiatives—a dramatic increase since 2012 when just 190 organizations were involved (GreenBiz.com). The following case study examples focus on those companies using natural capital to inform how they make and market greener brands.

The Coca-Cola Company—Return on Investment in Water Conservation

The world's largest beverage company has set ambitious water stewardship goals: protect water sources, reduce water use, treat all process water, and replenish all process water used, with the goal to be "water neutral" by 2020. The company developed a framework for more objectively evaluating the benefits of conservation investments and applied it to eight European replenishment projects. The framework included a measure of total ecosystem change, the net present value of the investment, and a metric called total investment multiplier to show how well the money was spent. The study identified project features and attributes that create greater return on investment (e.g., a large footprint coupled with low financial costs) and also concluded that stakeholder values are also an important consideration for the replenishment projects (Coca Cola 2016).

Ecolab—Water Risk Management and Shadow Pricing

Ecolab, the $14 billion global leader in water, hygiene, and energy technologies and services, wanted to find a way to help companies understand the financial risks they are facing from physical water scarcity and quality risks resulting from pollution. Working in collaboration with Trucost, the Water Risk Monetizer was launched in 2014 as the first tool to help companies assess the financial implications of local market failures to price water according to its availability, based on information about their future water use and production.

As water scarcity increases around the world, business leaders need actionable information to help them understand and manage their current and future water-related risks. The Water Risk Monetizer helps businesses make informed decisions to enable growth in this new era of water scarcity. (Emilio Tenuta, Vice President Corporate Sustainability, Ecolab)

The Water Risk Monetizer (www.waterriskmonetizer.com) combines an economic model on industry water use and treatment costs with scientific data on local water availability to calculate a water risk premium—or environmental shadow price—to better reflect the value a business should place on water.

A unique aspect of the tool is the way it incorporates a management framework for acting on both water quality and quantity risks, based on the potential for increased operating costs or revenue at risk. The tool also applies a context-based water reduction target, by estimating the amount of water available to a business—its "share" of total water available to all businesses in a water basin—based on the facility's contribution to the local economy (Figure 6.8). Because water is a shared resource among many users in a basin, it is essential for a business to understand how its allocation

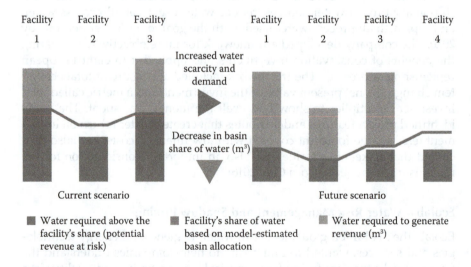

FIGURE 6.8
Estimating business value (revenue) at risk with Shadow Water Prices. (Courtesy of Trucost, London, UK.)

may change, and with it any potential for constraints to continued revenue growth. Companies use the tool to stress test business models, identify high-risk assets or regions, and build the business case for water efficiency where it most matters.

Eileen Fisher—Greener Product Development and Marketing

Eileen Fisher, a women's clothing and accessories company with social consciousness at the heart of its brand, has demonstrated leadership and made headlines in the apparel industry with its public commitment to source 100% organic cotton and linen by 2020. By applying natural capital valuations, Eileen Fisher determined its organically sourced cotton has natural capital cost benefits 2.5 times greater than conventionally grown cotton, providing science-based rationale and affirming Eileen Fisher's strategy to source organic fibers.

That less than 1% of global cotton production is organic creates familiar challenges for companies such as Eileen Fisher that set ambitious environmental goals: When supply is limited, prices go up. "As a company we ask ourselves, how long will this style last? Is the fabric sustainable? Does it cost too much? Will it sell? It's this constant balance between making things work for the business, the customer, and the environment."

At the time of the analysis, Eileen Fisher had already outstripped the market, achieving 69% organic fiber across its clothing and accessory lines. Eileen Fisher is actively informing customers about the natural capital benefits of organically grown cotton. For example, its website notes that while the initial cost of organic cotton is more expensive than its conventionally grown counterpart, consumers are making a long-term investment in the environment and human health.

The company is also taking significant steps to minimize other hot-spots of natural capital cost identified by Trucost's analysis. For example, it is addressing the natural capital cost of land use in its supply chain, accounting for its largest natural capital cost by far, by engaging internal stakeholders and suppliers to ensure its clothing is free from fiber derived from endangered forests and illegal logging. It is also acting to lower supply-chain greenhouse gas emissions by reducing its reliance on air shipping. And naturally, these messages are front and center of the company's customer communications, "our mission drives our business and our profitability fosters our mission." (Trucost 2016)

Green Electronics Council—Circular Economy and Greener Product Development

The Green Electronics Council used natural capital valuation to assess the business case for circular economy practices throughout the electronics sector. The results (see Figure 6.9) demonstrated that the industry could achieve a

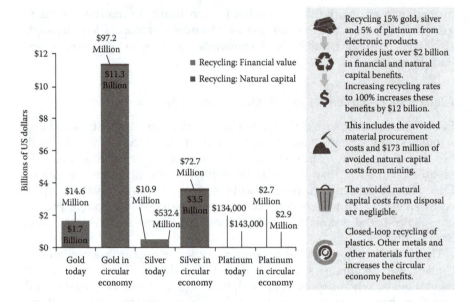

FIGURE 6.9
The benefit of Greener Electronics and a Circular Economy. (Courtesy of Trucost, London, UK.)

further $10 billion financial and natural capital cost-savings by increasing its recovery of gold, silver, and platinum from current rates to 100%.

Louis Vuitton Moet Hennessy (LVMH)—Shadow Water Prices and Supply-Chain Risk

Louis Vuitton Moet Hennessy, the luxury products group, was concerned that unknown exposure to water scarcity would affect its ability to operate and could lead to future impacts on profitability. The company developed a region-specific water risk map (Figure 6.10) across its entire value chain and identified regionalized natural capital shadow prices, taking into account local water availability. The regulatory risk framework combined an evaluation of shadow water costs and the local regulatory climate across LVMH's key sites and raw material inputs. By taking into account both the cost and likelihood of any risk, the framework allows LVMH to prioritize risk management where it most matters and strengthen the business case for investment.

Dell—Closed-Loop Recycling

Global technology leader Dell is seizing the circular business opportunity by integrating recycled plastic into the design of its OptiPlex 3030 All-in-One desktop computer. What is unique from other recycling initiatives is that the recycled plastic comes from used electronic equipment recovered through Dell's own global take-back scheme.

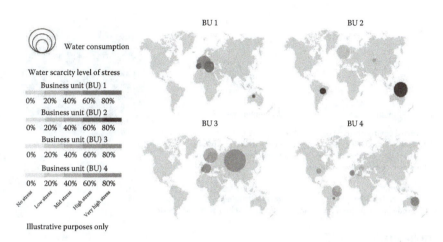

FIGURE 6.10
Water risk valuation based on local water use and scarcity. (Courtesy of Trucost, London, UK.)

Dell wanted to demonstrate the huge potential benefit of scaling up closed-loop plastics recycling and assessed the net environmental benefit of closed-loop recycled plastic in terms of lower pollution, reduced greenhouse gas emissions, and improved human health compared to using traditional plastic. This involved quantifying positive and negative environmental impacts and putting a monetary value—natural capital cost—on the result.

The results show that Dell's current usage of closed-loop plastic has a 44% greater environmental benefit compared to virgin plastic, equivalent to an annual saving to society of $1.3 million in avoided environmental costs. Of critical importance are the reduced human health and eco-toxicity impacts achieved by closed-loop recycling of plastic instead of disposal.

If all of Dell's plastic was supplied by closed-loop recycling, the environmental benefit to society would increase to $50 million per year (see Figure 6.11). If the entire computer manufacturing industry switched to using closed-loop recycled plastic, the environmental benefit would increase to $700 million per year (Dell 2015).

There are significant benefits to embracing a circular economy. Our closed loop plastics supply chain enables a resource-efficient product made from recycled content that costs Dell less. Companies need to realize sustainability programs are just good business. (Scott O'Connell, Director, Environmental Affairs at Dell)

Dell's net benefit analysis helps make the case for increasing the use of closed-loop recycled plastic both within its own business and across the industry. By switching to a more environmentally beneficial way

$1 million per year

Avoided environmental costs
to society from Dell's current
use of closed-loop plastic and
increased computer recycling

$50 million per year

Avoided environmental costs
to society if all Dell's plastic
was from closed-loop system

Assumptions:
35% closed-loop plastic, compared to 100% virgin plastic
Added benefits from increasing computer recycling, to recover plastic

FIGURE 6.11
Net benefit of closed-loop recycling. (Courtesy of Trucost, London, UK.)

of manufacturing products in which plastic waste is valued as a useful resource, Dell and the computer industry can take a big leap forward toward sustainability.

Kering—Greener Product Design Tools

Kering, the French luxury goods holding company, which owns Alexander McQueen, Balenciaga, Gucci, Puma, Volcom, and other sport and lifestyle brands developed the revolutionary concept for an Environmental Profit and Loss framework in 2011 and has since applied it across the operations and supply chain of its enterprise. In 2016, the company in collaboration with the Parsons School of Design created the *My EP&L* app that incorporates the EP&L framework into a simple, easy-to-use tool to measure and better understand the environmental impacts of students' creations. For example, *My EP&L* shows that by choosing a bag made from French leather with the inner lining in Chinese silk and hardware in brass from Chile instead of a bag made out of US leather with the inner lining in Chinese linen and hardware in Chinese bamboo, there is €4.40 less EP&L impact, or a 26% environmental saving (Kering 2016).

Conclusion

Quantifying the unpaid cost of environmental resource impacts and dependencies from pollution to natural resources and integrating location-specific shadow prices in decision-making provides vital insights that can help companies manage natural capital costs that are not yet fully priced by the market.

A growing number of companies are adopting this approach and putting shadow prices on carbon and other pollution emissions, water use, land conversion, and other natural resources to get ahead of the trend toward regulators, society, and nature correcting market failures to price-constrained environmental resources.

For many, it is a risk-management strategy—a way to identify high-impact environmental issues, assets, regions, and value-chain segments, and talk about risk in the language of business: dollars and cents. For others, it is a way to build the business case for energy efficiency, water conservation, and sustainable resource management where it most matters.

Leading companies are taking the insights a step further and using natural capital data insights to inform business decisions along the value chain in order to identify resilient strategies and innovative transformations that best position their business models for a low carbon, resource efficient future—and demonstrate the shared value they are creating for stakeholders and customers.

References

American Chemistry Council. 2016. Plastics and Sustainability: A Valuation of Environmental Benefits, Costs, and Opportunities for Continuous Improvement. https://plastics.americanchemistry.com/Study-from-Trucost-Finds-Plastics-Reduce-Environmental-Costs/ (Accessed December 29, 2016).

American Forest and Paper Association. 2016. http://www.afandpa.org/our-industry/economic-impact (Accessed December 29, 2016).

Brander, Luke. 2013. *Guidance Manual on Value Transfer Methods for Ecosystem Services.* UNEP. http://apps.unep.org/redirect.php?file=/publications/pmtdocuments/-Guidance%20manual%20on%20value%20transfer%20methods%20for%20ecosystem%20services-2013UNEP%202013%20Guidance%20manual%20on%20value%20transfer%20methods%20for%20ecosystem%20services.pdf (Accessed December 29, 2016).

Coca Cola. 2016. Natural Capital Accounting, the Coca Cola Water Replenishment Project. http://naturalcapitalcoalition.org/wp-content/uploads/2016/07/Denkstatt_Natural_Capital_Accounting.pdf (Accessed December 29, 2016).

Dell. 2015. Valuing the Net Benefit of Dell's More Sustainable Plastic Use. http://i.dell.com/sites/content/corporate/corp-comm/en/Documents/circular-economy-net-benefits.pdf (Accessed December 29, 2016).

Dow. 2016. 2025 Sustainability Goals. http://www.dow.com/en-us/science-and-sustainability/2025-sustainability-goals/valuing-nature (Accessed December 29, 2016).

Ecosystem Valuation. 2000. http://www.ecosystemvaluation.org/ (Accessed December 29, 2016).

GreenBiz.com. *State of Green Business Report, 2017.* Participation in Natural Capital Initiatives. GreenBiz.com. https://www.greenbiz.com/report/state-green-business-2017 (Accessed December 29, 2016).

International Integrated Reporting Council. The International IR Framework. 2016. http://integratedreporting.org/wp-content/uploads/2015/03/13-12-08-THE-INTERNATIONAL-IR-FRAMEWORK-2-1.pdf (Accessed December 29, 2016).

Kering Group EP&L. 2015. http://www.kering.com/en/sustainability/epl (Accessed December 29, 2016).

Kering, MyEP&L App. 2016. http://www.kering.com/en/press-releases/kering_and_parsons_school_of_design_collaborate_on_measuring_environmental_impact_of (Accessed December 29, 2016).

Natural Capital at Risk—The Top 100 Externalities of Business. 2013. *TEEB*. http://naturalcapitalcoalition.org/wp-content/uploads/2016/08/Trucost-Nat-Cap-at-Risk-Final-Report-web.pdf (Accessed December 29, 2016).

Natural Capital Coalition Protocol. 2016. http://naturalcapitalcoalition.org/protocol/development/ (Accessed December 29, 2016).

Puma EP&L. 2011. http://about.puma.com/en/sustainability/environment/environmental-profit-and-loss-account (Accessed December 29, 2016).

Stuchtey, Martin R., Per-Anders Enkvist, and Klaus Zumwinkel. 2016. *A Good Disruption, Redefining Growth for the Twenty First Century*. Bloomsbury, London.

TD Bank. 2015. Natural Capital Valuations. https://www.td.com/document/PDF/corporateresponsibility/2015-Natural-Capital-Valuations.pdf (Accessed December 29, 2016).

Trucost.. 2016. http://www.trucost.com/publication/growing-business-value-environmentally-challenged-economy/ (Accessed December 29, 2016).

Wealth Accounting Valuation System (WAVES). 2016. https://www.wavespartnership.org/ (Accessed December 29, 2016).

World Economic Forum. 2016. 2016 Global Risk Report. https://www.weforum.org/reports/the-global-risks-report-2016 (Accessed December 29, 2016).

Yarra Valley Integrated Profit and Loss Report. 2016. https://www.trucost.com/publication/yarra-valley-water-integrated-profit-loss-ipl-report/ (Accessed December 29, 2016).

Section III

Marketing Greener Products

7

Green Marketing

The Case for Green Marketing

Green marketing is quickly becoming more of a mainstream necessity rather than an initiative to be taken by proactive companies or those with a niche green product line. We are seeing green marketing occurring in all types of products and industries. It's not just the firms that sell direct to consumers that have to be concerned with this new marketing; suppliers of all types of products must start heeding this macro trend that doesn't seem to be subsiding, even during a global economic slowdown. Companies are seeing a need to develop greener products in all parts of business; consumer products, chemicals, electrical products, medical products, and even pharmaceuticals.

In working for a company that sells products directly to consumers and also does a good deal of business-to-business (B2B) sales, my experience tells me that green marketing is quickly moving toward being an imperative. I have seen green marketing take off in the consumer area, but I am now seeing an uptick in the B2B space too. Customers in all markets are looking for greener products. The signals in the field are building, and now is the time to start positioning your brand to get on board with this trend.

Consumer Demand

Global consumers are demanding products that have greener attributes; a Nielsen report indicated that "Fifty-five percent of consumers from 60 countries around the globe say they're willing to pay more for products and services from companies committed to making a positive social and environmental impact." Over 50% of consumers indicated they have purchased from companies that are offering products with environmental and social improvements and that before making a purchase they "check labels to see if a company is committed socially and environmentally." (Earth911 2016)

Fifty-five percent of consumers from 60 countries around the globe say they're willing to pay more for products and services from companies committed to making a positive social and environmental impact. (Earth911 2016)

We have seen greener products entering the mainstream in a major way. In the United States, when you make a trip to the local supermarket you will see products that use to only be offered in health food stores. Products like Seventh Generation detergents, Method facial cleansers, and Stonyfield Farm organic yogurt, to name a few, are commonplace. It is typical for grocery stores to have an "organic" section offering cereals, eggs, fruits, vegetables, milk, and baby products. On top of this, we see traditional companies coming out with green products that have done extremely well. Clorox's Green Works line is perhaps the best example of this. According to Clorox's own press release, 1 year after introducing Green Works, the "natural cleaning category has grown more than 100 percent, delivering on the company's goal to make natural cleaning more accessible and affordable to everyday consumers. Green Works™ is now the #1 natural cleaning brand in the United States with a 42 percent share of the market to date" (The Clorox Company 2010).

Should You Market Green?

What good is a greener product if no one knows about it?!

Further signals that emphasize the greener product trend is that consumers, despite high unemployment rates and low wages during the long economic slowdown, are willing to spend more for products that are environmentally friendly. In just a one-year period, "millennials willing to pay more for products and services from companies committed to positive environmental and social change increased from 55% in 2014 to 72% in 2015" (Georgetown Environmental Law Review 2016).

Another proof point for the growing desire of consumers to have products that are perceived as better for them is that organic products have reached a tipping point. The organic industry says US sales of its products "jumped 11 percent last year alone, to more than $39 billion." Also the number of US organic operations has grown by 250% since the government started certifying organic products in 2002, according to the Agriculture Department.

It is estimated that organics now make up almost 5% of total food sales in the United States. This growth is not just limited to food; according to the Organic Trade Association textiles, personal care items, and other nonfood organic product sales rose almost 14% last year and totaled more than $3 billion (Agweb 2016). Businesses are responding to these global trends as we can see by the actions of a variety of large companies.

Top 10 Big Business Response, to Meet Consumer Demand for Greener Products

1. *Panera Bread* "No No List": a list of more than 150 artificial preservatives, sweeteners, colors, and flavors that it plans to remove from its menu.
2. *Chipotle* becomes the first fast food chain to vow to remove genetically modified food (GMOs) from its menu.
3. *McDonald's* announced it will only buy and sell "chicken raised without antibiotics that are important to human medicine." It also announced it will offer milk from cows that are not treated with the artificial growth hormone recombinant bovine growth hormone in their US restaurants later this year.
4. *Tyson Foods*, the largest US poultry producer, announced plans to eliminate the use of human antibiotics in its chickens by September 2017.
5. Ice cream giant *Breyers* pledges to stop using milk from cows treated with recombinant bovine growth hormone, which has been linked to a number of health problems in humans.
6. *Shake Shack*, the owner of Panera, Five Guys, and Chipotle, announces the value of its initial public offering had increased to $675 million. They have a focus on healthier, ethical, and more sustainable food experiences.
7. *Home Depot and Lowe's*, the largest home improvement retailers, will phase out phthalates from vinyl flooring.
8. *Lowe's* commits to stop selling bee-killing pesticides (neonicotinoid) to protect pollinators.
9. *Adidas* announced plans to turn ocean plastic into sportswear materials made from ocean plastic waste.
10. *Levi's* and other companies commit to water conservation in drought-stricken California (EcoWatch 2016).

Sustainable Brands Are Profitable

In her book Green Giants, Freya Williams, the co-founder of OgilvyEarth, the sustainability communication firm, cites data that companies which lead in sustainable, social, and governance have **25% higher stock value** than their competitors and grow at double the rate than other brands. She goes on to make the case that greener products are profitable and highlights nine companies with over $1 billion in sales that are attributed to more sustainable products. As we can see from the list of these companies; greener products cut across various sectors and are not only better for people and the planet but also are yielding significant profit. This is another reason why every company should be making and marketing greener products seriously.

The list below indicates nine companies that have businesses with over a billion US dollars in annual revenue connected to a product or service with "sustainability at its core." It should be noted that this is not a comprehensive list because there are other companies that have over a billion in revenue from greener products that are not included here. For one, the company I work for, Johnson & Johnson, has at the writing of this book, $11.5 billion in Earthwards® recognized greener products on the market.

Companies with over $1 Billion in Sales of Greener Products		
Brand	**Sector**	**2014 Revenue in Billions ($)**
Nike Flyknit	Athletic Apparel	1
IKEA (products for a more sustainable life)	Furniture	1.13
Natura	Consumer Products	2.65
Tesla	Automotive	3.2
Chipotle	Food	4.11
Whole Foods Market	Food	14.19
Toyota Prius	Automotive	15.44
GE Ecomagination	Diverse Products	28
Unilever	Consumer Products	52.37

Greener Products Should Not Command a Higher Price

This idea of **greener being an** "and" is a concept I believe is proven out by numerous studies of consumers' purchase tendencies. What I mean by this is that the majority of consumers want a greener product, but don't want to pay more for it so it is an "**and**." Only the deep green purchasers, or the ones the Shelton Group would call "Actives" would be willing to pay a premium for a greener product (only 22% of shoppers in the United States) (Iannuzzi 2011). Therefore, in order to reach the majority of purchasers, product attributes like efficacy and appeal must be present and then its sustainability qualities can push it over the edge. Consumer product buyers are not the only ones looking for greener products; we are seeing companies having to market their products' greener aspects to other businesses to get market advantage too. The concept of the greener characteristics being an "and" also applies to the business customer as well.

Business-to-Business Demand

Similar to consumers, many companies are actively seeking to purchase greener products. Firms have been pressured by market forces to take responsibility for the supply chain. Perhaps the most notable example of this is when Nike and other footwear and apparel companies had received significant pressure to take responsibility for the poor working and environmental conditions at the companies that manufactured shoes for them in the 1990s. Not long after this, several apparel firms had received significant pressure to take responsibility for suppliers that made their products in sweatshops. Companies started to get the point and then began to develop very comprehensive policies and auditing programs for their suppliers to insure that the environment is protected and employees within their supply chain have safe working conditions. Taking this initiative even further, we see manufacturing firms now asking their supply chain to achieve certain goals like reducing water, energy, waste, and other issues. These initiatives have been coined as "**greening the supply chain.**"

With the advent of greening the supply-chain initiatives, suppliers will gain market advantage when they meet the demands of their business customers. I know that in my company, we give preference to like-minded firms that have adopted more sustainable practices. Meeting your customer's needs is the inspiration behind B2B marketing.

As an illustration, look at one of Unilever's sustainability goals: to increase the amount of sustainably sourced raw materials such as palm oil or tea (Unilever 2016). If your company is a supplier of one of their raw materials and you can reliably source sustainable tea, you can help them with this goal and it will give you an opportunity to increase your sales.

Similarly, suppliers to Tesco, the largest retailer in the United Kingdom, can see that they are very serious about getting sustainable innovation from their suppliers. They have a comprehensive Environmental Guide for Suppliers and established targets to drive improved performance. Being a major global retailer, Tesco recognizes that they have opportunity to make a big impact in their supply chain by conducting their operations in a responsible way. Strides have been made to reduce their carbon footprint, and they want to encourage their supply base to do the same. A goal was established to become a zero carbon business by 2050 and to reduce CO_2 emissions 30% from the products in their supply chain against a 2008 baseline by 2020. If you sell to Tesco, you will have to put plans in place to address CO_2 since this goal affects all suppliers.

Addressing carbon emissions isn't the only concern for suppliers; Tesco identified climate change, water use, and biodiversity as key areas for their suppliers to focus on. They believe that in making strides to reduce their footprint, targets need to be established for their biggest agricultural products by 2025, specifically:

- 30% reduction in greenhouse gas emissions
- Reduction in water use, including local reduction targets for water-stressed areas
- Improvement in farmland biodiversity (soil health, pollinators, and off-field biodiversity)

Tesco goes even further for certain suppliers of agricultural commodities because poor practices from farming have been associated with deforestation. A goal of Zero Net Deforestation was set for 2020. The key commodities being focused on are palm oil, cattle products, soy, and timber. In addition, mandatory policies for commodities of high concern have been established for suppliers to follow, for example, palm oil sourcing and sustainable sea food. Any company that can excel in these areas will at a minimum get the right to sell in a Tesco store and, if they have superior performance, could get the most preferred shelf space or other marketing benefits (Tesco 2016).

It's not just Tesco that has green purchasing policies. You would be hard-pressed to find a large company that doesn't have a procurement strategy that emphasizes purchasing greener products. Focus has been brought on suppliers in various product categories like energy savings, and electronics, to paper that contain post-consumer recycled (PCR) content or greening the supply-chain initiatives that seek suppliers to produce sustainable innovations. These kinds of initiatives will only become broader and demand more eco-efficiency from suppliers in the future. The companies that see this trend and can meet customers' demands will be able to win in the greener product market.

Supplier Scorecards

B2B marketing also has a big impact and one of the reasons is the advent of supplier scorecards. Walmart's Supplier Sustainability Assessment and Packaging Score Card is well known, but there are other very influential companies with sustainability scorecards of their own. Large companies like Kaiser Permanente and Procter & Gamble (P&G) have also issued mandatory supplier scorecards. Whatever market my company is selling product to, be it retailers or hospitals, there are always sustainability questions associated with requests for proposal and tender offerings. I have heard customers say that up to 20% of the purchasing decisions can be based on sustainability performance. Suppliers take notice when they hear the VP and Chief

Procurement Officer of a major health-care provider, Kaiser Permanente, say "green up your act today, lest you lose a huge client tomorrow." Also consider that Kaiser's purchasing comes to about $14 billion a year and P&G alone has 75,000 suppliers throughout the world (Guevarra 2016).

These scorecards will no doubt drive companies to highlight the greener benefits of their products. When customers are interested in whether or not the products they are buying contain specific toxic chemicals, or if they are manufactured using renewable power, it is in your best interest to clearly communicate which products in your portfolio have greener features.

Examples of Questions from P&G and KP Scorecards

- What % of energy consumed is generated from renewable resources?
- Does the company have a climate action plan with baseline and targets?
- How many metric tons of hazardous and nonhazardous waste are produced? (Joseph 2010)
- Free of intentionally added bisphenol A or bisphenol A derived chemicals (including thermal paper)?
- Free of polyvinyl chloride?
- Primary Packaging—Contains more than 10% PCR content? (Practice Greenhealth 2016)

Other drivers of B2B marketing are the public relations benefits for being perceived as a company that cares, and helping customers save money. Home-improvement giant Lowe's was recognized as the WaterSense® retail partner of the year by the U.S. Environmental Protection Agency. They also received kudos from the Department of Energy's ENERGY STAR® Sustained Excellence Award. These accolades were achieved by providing products that save money while reducing environmental impact.

In 2009, Lowe's sold enough ENERGY STAR products to save consumers more than $265 million each year off their energy bills compared with non-ENERGY STAR qualified products. Lowe's explains the benefits of water conservation in their stores and helps families reduce utility bills. The number of WaterSense-labeled toilets and bathroom faucets, Lowe's sold in 2009, alone could save consumers $13 million each year on water bills (Lowe's 2010).

With this kind of focus on greener products, would it make sense for suppliers to get the ENERGY STAR or WaterSense label on their products? With Lowe's getting such positive recognition as a sustainable company while generating savings for their customers, it would be prudent for suppliers to provide products that help Lowe's attain their goals.

This scenario is common across all industrial sectors. Companies that provide building supply products can help their customers seeking green building certifications (like LEED) with products such as low VOC paint, more energy-efficient windows, sustainable wood products, more efficient air-handling equipment, more sustainable carpet and cleaning compounds, to name a few. We are seeing that suppliers are being held to higher standards, sustainable innovation is one of them. When a company can provide greener products to their business customers and help them achieve their sustainability goals, they are more likely to make the sale.

Examples of Green Marketing

Earlier in this book, we discussed the practices leading companies use to make greener products. **Step one in green marketing is to have a credible greener product** to bring to customers. The next, and just as important step, is how to appropriately communicate the qualities of your product or service that meets customers' demands.

What are the key elements of green marketing? First, your product has to have a "greener" story to tell based on scientific facts and data. You must also understand the market segment you are selling into and seek to meet your customer's greener product demands. Finally, the greener benefits must be appropriately communicated, without overstating or misleading.

The Boston Center for Corporate Citizenship prescribes five guidelines for green marketing.

1. *Be precise*—Make specific claims that provide quantitative impacts.
2. *Be relevant*—Demonstrate a clear connection between the product or service and the environment.
3. *Be a resource*—Provide additional information for consumers in a place where they want it.
4. *Be consistent*—Don't let marketing images send a signal that contradicts the carefully chosen words and facts you use.
5. *Be realistic*—There are always more environmental improvements that can be made to a product or service, and improvements are but one piece of a much larger environmental journey (Hollender et al. 2010). The way I would say this is, communicate your products as green*ER*, not green.

These five guidelines make a lot of sense to me. Based on my experience, I would simplify effective green marketing into three key elements; first, there must be a truly greener product to market (see the chapter on making greener products to know how); second, we must understand what the customer requirements are; and third, appropriately communicating greener characteristics is vital.

Keys to Green Marketing

- Have a credible greener product story
- Meet your customers greener product demands
- Appropriately communicate the product's greener attributes

Business-to-Consumer (B2C) Examples

Green marketing is all about communication. The greenest product in the world is useless if no one knows it's available. We have many successful green marketing campaigns to learn from, so let's evaluate a few and see what the key aspects are looking through the lens of:

1. Have a credible greener product story,
2. Meet your customers greener product demands, and
3. Appropriately communicate the product's greener attributes.

Clorox Green Works

As we previously discussed, Clorox Green Works changed the game and enabled greener products to be sold in the mainstream market. Prior to Green Works, we have seen green products stigmatized in the consumer mind as being a niche product sold primarily in health food stores. The mainstream consumer has impressions of these products as not working well, being too expensive, and that they couldn't be trusted. Green Works is based on a natural ingredients platform (at least 99% natural ingredients). In their testing and preliminary evaluations, they determined that Green Works cleaners performed as well and sometimes better than cleaning products that were on the market. Consumer research indicated that the highest-scoring desire was that of personal protection: "doing things that protect me and my family."

To address the possibility of not being considered authentic, the line of products was put through the U.S. EPA's Design for the Environment (DfE) certification (now called Safer Choice). Further bolstering their green claims when they first introduced Green Works, they received endorsement from the environmental group the Sierra Club, placing their logo on product bottles. Augmenting this, they placed the ingredients of the compounds used on the product labels, even though this was not required by law and is not a common practice for household cleaners (Werbach 2009).

Green Works clearly has done the work to credibly develop a greener product. They understand consumers' needs for personal protection, and they have clearly messaged the products greener attributes to the customer. Receiving third-party endorsement with the DfE certification and the Sierra Club's backing helped cement the product as truly green in the mind of prospective purchasers.

Seventh Generation

Coming from a different direction than Clorox, Seventh Generation is a deep green company from its very beginning, also selling cleaning products, but their offerings include baby, laundry, dish detergents, and other products as well. Their aim was to bring their greener products into the mainstream market. In their mission, they state an Iroquois Indian law that they must "consider the impact of their decisions on the next seven generations." You can't make a stronger commitment to sustainability than that in your mission statement.

Consumers purchase all-natural personal care products to limit exposure to toxins and chemicals—this is the essence of Seventh Generation products. They have been effective on communicating that their products are greener. Their website contains a wealth of sustainability information; products are based on natural ingredients, and all products carry eco-labels; for example, USDA Certified Biobased, cruelty-free (Leaping Bunny), How to Recycle logo, and some even are labeled as gluten free. They are very clear about their sourcing policies, indicating supplier requirements, and even the location of manufacturing sites—important issues for conscious consumers. I really like how they have an ingredients "glossary" where they indicate every material used in their products with an explanation of its purpose and its environmental impact. Customers can use this to look up what an ingredient is used for and if it is safe for the environment. As an example:

> **Ingredient:** *Aloe barbadensis* leaf juice
>
> **Use**: Skin Conditioner
>
> **Environmental Impact**: Plant derived, biodegradable

Let's take a look at the messaging for the Baby Shampoo & Wash: A natural, USDA Certified Biobased product, 96%. Gentle, foaming baby shampoo and wash. Tear-free; lightly scented; massage foam into baby's skin and hair, then rinse well; safe + effective formula; made with pure coconut oil + other plant-based ingredients; no sulfates, parabens, phthalates, or formaldehyde donors. We care about your baby AND the planet. We're committed to sustainable ingredient choices, such as organic coconut oil, that help protect forest habitats and the animals that call them home—like the adorable playful orangutan you see on the front of this package.

Eco-labels used are USDA Certified Biobased, cruelty free, how to recycle info, gluten free. A full disclosure of all ingredients are noted on the website.

Communicating Green Attributes is Important to the Customer

Seventh Generation Baby Shampoo & Wash	
Messaging	
Natural	USDA Certified Biobased Product
Gentle	Foaming baby shampoo and wash, Tear-free, lightly scented
Safe + effective formula	Made with pure coconut oil + other plant-based ingredients; no sulfates, parabens, phthalates, or formaldehyde donors

Source: Seventh Generation, 2016, Baby Shampoo & Wash, http://www.seventhgeneration.com/baby-shampoo-wash?v=1116

WHAT'S INSIDE OUR SAFE AND EFFECTIVE FORMULA

Water, *Aloe barbadensis* leaf juice (plant-derived emollient), coco-glucoside (plant-derived surfactant), glycerin (plant-derived humectant), heptyl glucoside (plant-derived cleanser), sodium chloride (mineral-based thickener), organic *Cocos nucifera* (coconut) oil (plant-derived skin conditioner), citric acid (plant-derived pH adjuster), d-dodecalactone, delta-decalactone, dimethyl heptenal, gamma octalactone, hexyl acetate, *Jasminum sambac* (jasmine) flower extract, maltol, sweet orange peel oil, vanillin, glyceryl caprylate (plant-derived preservative), and caprylhydroxamic acid (synthetic preservative booster).

Product Manufactured in: South Carolina, USA

Ingredient Origins: USA and Globally Sourced

(Seventh Generation 2016)

Seventh Generation covers all the key areas for green marketing. They obviously have a truly greener product, they know what their customers are looking for (more natural ingredients), and they clearly communicate the greener benefits to meet customer demands, on pack as well as digitally.

Honest Tea

I first came across Honest Tea through my teenage children. Desiring a low-calorie, good-tasting drink, they became aware of this beverage. The name of the product describes their mission. "Honest Tea creates and promotes delicious, truly healthy, organic beverages. We strive to grow with the same honesty we use to craft our products, with sustainability and great taste for all."

A commitment to social responsibility is central to Honest Tea's identity and purpose. The company states that they strive for "authenticity, integrity and purity, in our products and in the way we do business." The platform for products is a healthy beverage with a lot less sugar than most bottled drinks. In March 2011, The Coca-Cola Company purchased Honest Tea, however, it is run as an independent business unit. An evaluation of their website certainly reinforces that the business unit is operating independently.

There are Five Pillars to their Mission:

- Promoting Health & Wellness
- Reducing Our Environmental Footprint
- Creating Economic Opportunity
- Maintaining Transparency
- Building communi-Tea (Honest Tea 2016a)

Independent laboratory analysis is used to prove that its drinks have antioxidant levels that are as high as or higher than brewed tea leaves. In 2003, Honest Tea became the first to make a Fair Trade Certified™ bottled tea. This certification strives to empower family farmers and workers around the world to get a fair price for their harvest, have safe working conditions, and earn a living wage (Honest Tea 2011).

Concerned with unhealthy super-sweet beverages loaded with sugar, Honest Tea decided to make beverages without any sugar such as "Just" Green Tea (no calories, no sugar). They also produce products that are a "tad sweet," drinks sweetened with organic cane sugar which contain 60 calories or fewer—far less than the standard bottled iced teas on the market. Heavenly Lemon Tulsi Herbal Tea is an example—"just a tad sweet" brew made with lemon juice, tulsi, and rooibos leaves, and an aromatic blend of lemongrass, lemon peel, and lemon myrtle, which has 60 calories.

Honest Tea beverages and tea bags are certified to the USDA's organic standards. The certification insures customers that the products' raw materials were grown following organic farming techniques and do not contain antibiotics, pesticides, irradiation, or bioengineering. The farms that provide the raw materials are examined by third-party certification agencies.

Honest Tea Messaging to the Health-Conscious Consumer

All of our teas are certified organic according to standards set by the US Department of Agriculture (USDA) and enforced by accredited third-party certifying agents.

Glass, PET plastic, boxes, and pouches for children's drink containers are used for packaging. Improvements have been made to pouches, which are typically not recyclable. To address difficulties with recycling pouches, their new aseptic Tetra Brik® package is recyclable in over 50% of US municipalities. Because most drink pouches cannot be recycled in curbside programs, a partnership was made with the company TerraCycle, which upcycles these packages and uniquely converts them into useful items like fashion bags, tote bags, pencil cases, and other items. In 2014, there were over 235 million pouches recycled by TerraCycle. This type of innovative program gives customers confidence in Honest Tea because it conveys that they are doing their best to facilitate the reuse of their packaging.

Communicating the greener benefits of their products is done very clearly on bottle labels. As an example, the Fair Trade Certified and USDA Organic logos along with the calories (60) are prominent on the label of their Peach OO-LA-Long "Just a Tad Sweet" tea. A visit to their website reveals further messaging that is focused on the conscious consumer by using the following images: Fair Trade Certified, USDA Organic, Gluten Free, no GMOs, and 1% for the Planet. There is even a source map that depicts which parts of the world key ingredients such as tea, fruit, herbs, spices, and sweeteners are sourced from (Honest Tea 2016).

Certifications and Claims used by Honest Tea

- Fair Trade Certified™
- USDA Organic
- Calories per bottle
- Gluten Free
- No GMOs
- 1% for the Planet

To further reinforce their commitment to sustainability, they have developed a "Mission Report." In the 2015 "Honest Mission Report," there are facts, figures, data, and details that build consumer confidence that they are keeping their eye on the sustainability ball. An example is the discussion of their donations to 1% for the Planet. They give 1% of sales from their 16 fl. oz. Glass-bottle tea line to organizations within the 1% for the Planet network. There is also an interesting discussion about their sourcing practices, where fair trade premiums are paid to suppliers for efforts to help make their practices more efficient so that they can garner more income.

Honest Tea has a good sustainability story to tell that is backed up with data and third-party certifications. Research indicates that the primary driver for organic food purchases is the desire for "better health." Consumers tend to favor endorsements or certifications for this category of products. Therefore, they are on target in addressing their customers' desires with the organic and fair trade certifications they obtained and prominently display. Finally, they are credibly and very clearly communicating to their customers the products' greener traits.

Timberland

Being one of the first companies to develop a product-specific scorecard called the Green Index®, Timberland has built sustainability into its core. Supplying footwear and apparel for the outdoors, they feel a direct connection to developing sustainable products—hence their environmental commitment is called "Earthkeeping."

Timberland uses the Green Index label to depict their environmental improvements—it looks similar to a nutrition label. This label indicates the climate impact, chemicals used, and resource consumption. The label conveys criteria that provide consumers with a relative measure of a product's environmental impact to spur more sustainable purchasing.

An example of product improvements is the Earthkeepers® GT Scramble Lace Chukka.

This boot scored 4.5 out of 10 on the Green Index (10 being the worst score). Their three categories scored as follows:

- 3.0 for Greenhouse gases which are produced in making the raw materials during footwear production.
- 2.0 for Chemicals used in materials and footwear production, and
- 7.5 for Resource Consumption for making the product. The idea is to use resources that use less land, water, and chemicals, and more recycled materials.

The product improvements are communicated through the Green Index label (Timberland GT Scramble 2016). Further demonstration that Timberland gets how to do green marketing right is the messaging for their Earthkeepers® original leather boot. Consumers looking for a boot that has been made with lower environmental impact can quickly see that thought was put into making this product greener (Green Index 2016).

Marketing Earthkeepers® Original Leather Boots

- We haven't sacrificed quality or rugged good looks to make them eco-conscious
- 100% recycled PET lining is made from recycled plastic bottles
- 100% organic cotton laces
- One or more major components use at least 50% leather tanned in a facility rated Silver or higher by a third-party environmental audit

(Timberland Earthkeepers® Original Leather Boots 2016)

Timberland has developed products using their Green Index that have better environmental performance. Customer demand for transparency about the materials used in their products are being met. Consumer research has been used to identify the most important desires to communicate about. And they have come up with a unique communication tool similar to a nutrition label to fulfill their customer's interest.

Neutrogena® Naturals

Neutrogena is an international consumer brand that includes facial products, hair care products, cosmetics, and products for the skin and is one of the Johnson & Johnson companies. To address marketplace demands for greener products, they released a new line called Neutrogena® Naturals products in 2011. A lot of good green marketing concepts have been put into this initiative, and it will be informative to evaluate their approach.

The product is positioned as having the best ingredients nature has to offer, "merging them with our clinically proven expertise to bring you the best of both worlds—naturally derived skincare that works. Neutrogena® Naturals products are designed to be effective with absolutely no tradeoffs." The products in this group include makeup remover cleaners, facial cleansers, and moisturizers and are based on a formulation containing 94% naturally derived ingredients (Figure 7.1).

All product packaging in this line has the same look and is based on a naturally derived ingredient platform. Some paper cartons are made from 100% recycled paper with up to 60% PCR content. Plastic bottles also have been greened up and contain up to 50% PCR content; a smart move because ecologically aware customers take notice of environmental issues with the packages for the products they buy.

The marketing clearly addresses a significant concern of consumers; that greener products do not work as well as their less environmentally conscious counterparts. On the product label and website, one of the key messages are: "Neutrogena® Naturals product is made according to our core values, with the same standards of efficacy and excellence as every Neutrogena® product you know and trust;" in other words they work! The desire of the natural product consumer to limit their exposure to toxins and hazardous chemicals is balanced with a message that this greener product comes from a brand that has been around awhile, one that you can "trust" (Neutrogena Naturals 2017).

The communications used to convey that the product's greener attributes are: contains naturally derived ingredients, "no harsh chemical sulfates, parabens, petrolatum, dyes or phthalates." On the product lines website, there is much more information on how they tie sustainability into the core of the brand. There are multiple references to the efficacy of the product line and the science to give consumers confidence in a naturally derived product; for example, statements like this are used: "leveraging our skincare

FIGURE 7.1
Neutrogena naturals product.

expertise, we've created pure, high-quality, naturally derived formulas that deliver results." The proof points for their sustainability story are described with six key elements, which will resonate with consumers that are interested in a natural-based product.

- *Naturally derived*—plant-derived ingredients
- *Not tested on animals*—no testing of cosmetic ingredients on animals
- *PCR plastic*—up to 50% PCR plastic used in bottles
- *100% recycled paper*—folding cartons are 100% recyclable and contain 60% PCR
- *Supports clean water*—partnered with the Nature Conservancy to protect sources of clean water (Neutrogena Naturals 2017)

The overarching message is similar to Clorox Green Works in that it comes from a company with a long history of safe and effective products. This messaging develops customers trust that natural products will meet the same standards as the base products line. Neutrogena has done a nice job of putting it all together. They have data to back up that their products and packaging are truly greener. The customer that is most likely to purchase a natural product is interested in reducing the amount of toxic substances they are exposed to; this product line credibly addresses this concern with their Naturals platform and greener packaging. The communication of their efforts is performed plainly on product labeling as well as digitally through their website. In addition, the relationship with the Nature Conservancy and the use of natural ingredients appeals to shoppers interested in health and wellness.

Neutrogena® Naturals
Fresh Cleansing Makeup Remover Label Messages

- No harsh chemical sulfates, parabens, petrolatum, dyes, or phthalates
- Completely removes makeup and gently cleanses for refreshingly clean skin
- Peruvian Tara Seed bionutrient-rich cleanser renews and refreshes

---

Business-to-Business (B2B) Green Marketing

When we think of green marketing, our natural tendency is to think about direct to consumer products. However, there is a far greater opportunity in B2B marketing. Consider the millions of suppliers that are necessary for all of the products in the marketplace. As mentioned above, P&G has 75,000 suppliers alone! I believe that in the B2B space, green marketing will be an imperative in the future, and for now it is a key product differentiator. So let's evaluate some successful B2B green marketing campaigns and see how they compare to the B2C approach.

Ecomagination

Perhaps one of the most successful B2B marketing initiatives is GE's Ecomagination program. It would be hard to discuss green marketing without mentioning the inroads made by GE. There are Ecomagination television commercials, print advertisements, and digital marketing as well as an annual report that details the programs' success. GE's CEO is seen at numerous events throughout the world effectively communicating about this program.

Ecomagination was launched in 2005 and has steadily grown into one of the most successful green marketing programs ever, paying dividends to GE and their customers. In fact, in 2015, sales of greener products resulted in $36 billion! (GE 2015). The whole reason for the existence of Ecomagination is to *meet customers' requirements*. The 2009 Ecomagination report states: "Ecomagination is a business initiative to help meet customers' demand for more energy-efficient products and to drive reliable growth for GE." A program can't get any closer to meeting customer's needs than that. In addition, if you polled people throughout the globe on what is the most important environmental issue facing the world, inevitably you would hear global warming or climate change. GE's program sets to tackle this issue through this business program.

Meeting Customer Demands with Greener Products

Ecomagination is a business initiative to help meet customers' demand for more energy-efficient products and to drive reliable growth for GE. (2009 Ecomagination Annual Report)

The program claims that there are many examples of greener products. The 2014 Ecomagination report boasts that they have some of the most efficient products in the world. Some examples include the Tier 4 Locomotive, which decreased emissions by approximately 70% or more over their Tier 3 product and saved customers an estimated $1.5 billion. The HA-turbine is the largest, most fuel-efficient gas turbine in the world, at more than 61% efficiency. The LEAP jet engine gives customers a 15% improvement in fuel efficiency versus its predecessor and also provides improvements in noise and emissions, and the lowest overall cost-of-ownership in the industry—a critical aspect to the airline industry. The city of San Diego employed Ecomagination's "intelligent" lighting system, called LightGrid, which links its streetlights to the Industrial Internet. The city replaced more than 3,000 light fixtures with GE LED lights, their intelligent street lighting systems can reduce electricity consumption by 50–70%. If you were in the market for one of these products, surely you would take a good look at Ecomagination products. The key messaging here is that it saves you money while providing significant environmental improvements.

Ecomagination has been an extremely successful green marketing program. They have developed a robust portfolio of greener products (see the case study in the Developing Greener Products chapter). The whole point of the program is to meet customers' needs. Marketing is positioned and connected right at the core requirements—more energy- and water-efficient and clean-power generating products. They have revolutionized the way green marketing communication is conducted, and the case could be made that GE has brought B2B green marketing mainstream due to its success and market penetration.

BASF

One industrial category you would probably not think is doing green marketing is chemicals. However, this may be one of the most beneficial categories to emphasize the greener aspects of your products. I recall seeing television commercials years ago by BASF, a German chemical company, saying, "We don't make a lot of the products you buy; we make a lot of the products you buy better." Well, today they can say, we make the products you buy "greener."

BASF's platform for communicating their more sustainable products is called Sustainable Solution Steering®. This process puts all products into four categories tied to providing sustainability solutions to their customers. The "accelerator" category is the highest level and is described as products that provide a "solution with a substantial sustainability contribution in the value chain." As a chemical company, they believe chemistry is an enabler offering "business opportunities" for meeting customer needs.

One of the customer groups that BASF services is automotive. Focusing on energy efficiency and reducing air pollution are key sustainability concerns for this industry. To address this need, BASF offers lightweight plastic materials which lower the weight of the cars, resulting in better fuel-efficiency and catalysts which reduce exhaust emissions to improve air quality.

Another category that they sell to is packaging. Here, the use of materials that are lighter weight or biodegradable are sustainability delighters for their customers. An accelerator product in this area would be the biodegradable ecovio® paper coating, which enhances the proper disposal of paper-based products. Another would be the water-based resin Joncryl FLX®, which is an alternative to solvent-based printing technologies, reducing volatile organic compounds emissions from the printing process on packaging and the overall packaging life cycle (BASF 2017).

It makes a lot of sense for chemical companies to focus on the sustainability needs of their customers since what they sell is the building block for all products. BASF is meeting the demands of customers through products that lower environmental impact such as lightweight plastics. Green claims are backed up by their internal sustainability rating system for all products, which also enables communication of the greener attributes to their customers.

Steelcase

Every business needs office furniture, and most companies nowadays have an environmentally preferred purchasing program that addresses office furniture and equipment. If you think about the most important issues from a sustainability perspective for office furnishing, you would probably come up with materials used and the end of life of the product. Businesses don't want materials of concern in their products like flame retardants, toxic metals, and wood that came from endangered forests. And when it's time to get new furniture, they don't want the old equipment to go into a landfill. Steelcase addresses these issues head on in their messaging for their products.

Steelcase is an over 100-year-old company headquartered in the United States that sells globally. Their product portfolio includes chairs, tables, bookcases, lighting, and screens—basically anything you need for your office to operate. The raw materials for these types of products include, plastics, metals, wood, and leather, all of which have their own set of environmental issues associated with them. Recognizing the importance of purchasing more sustainable office furnishings to their customers, Steelcase has "customers" at the center of their sustainability focus. Their vision is listed as follows:

> "Our sustainability vision is clear: bring lasting value to our customers, employees, shareholders, partners, communities and the environment."

To address the use of materials and end-of-life issues they have deployed design for the environment and circular economy thinking. "We design products for circularity by avoiding and eliminating materials of concern, optimizing performance throughout the life cycle and for remaking, recovery, and end of life strategies. We are actively working with our supply chain to eliminate and phase out materials of concern and to develop suitable alternatives where they may not yet exist" (Steelcase Sustainability 2017).

If you were in the market for new furniture, it would be great if the company you are buying is also responsible for managing your old stuff. In their 2015 sustainability report, Steelcase indicates that they executed over 1,300 customer requests to decommission furniture and claim to have saved their customers $2.3 million by reuse of the furniture. In addition to this, 6.4 million pounds of furniture was diverted from landfills and was recycled or sold as pre-owned (Promise + Progress 2016).

Steelcase End of Use Messaging

Our products are built to last; sometimes they outlast customer needs. Therefore, we offer multiple programs to extend a product's lifespan through reuse and recycling, refurbishing and donating. These end of use services keep furniture out of landfills, provide non-profit organizations with needed resources and help customers meet their sustainability goals. (Steelcase Sustainability 2017)

A really good practice employed includes the use of product environmental profiles. These documents tell a customer everything they need to know about key sustainability elements. Evaluation of the environmental profile of the Siento® office chair addresses issues that are important to customers such as certifications that give customers confidence regarding the absence of materials of concern and how it contributes to LEED certification. Below is part of the data that appear in the environmental profile.

Siento® Office Chair Environmental Profile Information

Certifications—Cradle to Cradle™ Certified—Silver (depending on options), SCS Indoor Advantage™ Gold certified for indoor air quality in North America, level® 3 certified to ANSI/BIFMA e3 standard

LEED Contribution—Recycled content, regional materials, rapidly renewable materials, low-emitting materials, sustainable purchasing, innovation in design

Environmental Facts—91% recyclable, 28% recycled content, life-cycle assessment completed, PVC-free and chrome-free, standard leather is chromium-free.

Reviewing this environmental profile gives customers confidence that sustainability issues have been well managed. It also makes it easy for customers to calculate the environmental improvements realized from their purchase by listing the percentage and total weight of recycled content and how they can get LEED points.

Steelcase is a very good example of B2B green marketing. They make customer concerns the center of their marketing program. There are many good methods deployed to entice customers to buy from their company. The use of third-party certifications, product environmental facts, providing data for the customer to easily calculate their environmental improvements through purchase of their products and providing end of life solutions are all great ways to sell products to the growing number of companies that want to purchase more sustainable office furniture and equipment.

Sodexo

Sodexo, Inc., the largest food service company in the world, has also embraced sustainability through their *Better Tomorrow Plan*. They have three core pillars to this plan:

1. *We are:* The fundamentals that are the cornerstone of a responsible company.
2. *We do:* Four priorities—a responsible employer, nutrition/health/ wellness, local communities, and the environment.
3. *We engage:* Dialog and joint actions with our stakeholders (Sodexo 2016).

One of their commitments, to source sustainable fish and seafood, has resulted in a good green marketing story. It is reported that seven of the top 10 marine fisheries are over-fished. Sodexo has pledged to make a "positive impact on the health of the world's oceans and fisheries by significantly reducing the amount of unsustainable seafood in the food service industry." To put teeth into their commitments, in 2011 they signed a global agreement with the Marine Stewardship Council (MSC), a nonprofit independent organization that certifies that wild-caught fish are not species which are at risk by overfishing. Sodexo also has partnered with World Wildlife Fund (WWF) and works with other nongovernmental organizations to be advised regarding responsible seafood sourcing. Traceability certification ensures its clients and consumers that MSC-certified products are not mixed or replaced at any stage of the supply chain with non-certified seafood (Sodexo report 2016). Making a commitment like this will give Sodexo a competitive advantage for customers that are interested in purchasing a sustainable product.

Having certified seafood will make it easier for Sodexo to sell to customers that are being pressured by environmental groups because of the overfishing that has occurred. Sodexo has developed a greener product to meet customer needs—a certified sustainable food. They have clearly committed to responsibly sourcing seafood and communicating their company's greener benefits. A partnership with an independent nongovernmental organization to certify their products sustainable benefits helps to solidify their products credentials and protects against green washing.

Sealed Air Diversey Care

Diversey Care is a business unit of the Sealed Air company, which manufactures and markets detergents, cleaners, sanitizers, lubricants, floor care products, carpet cleaners, and carpet cleaning and floor care machines, along with a host of other products and services. Their primary customers are hotels, hospitals, and companies that operate office and manufacturing buildings. Most companies discuss sustainability on their website as one of their objectives; however, Diversey has it integrated into their business model.

One of the keys to green marketing is to know your customers' needs and to help them with their sustainability goals. This commitment is plainly stated when describing their products; "at Diversey, we understand what it takes to run a successful and sustainable facility cleaning operation." If clients are asking for 'sustainable cleaning' or services that are 'environmentally friendly,' "Diversey can help you with minimizing the impact on the environment, not only with the products you use, but also in the way you conduct your daily business...." (Sealed Air Diversey 2017)

A perfect illustration of Diversey putting sustainability into practice is their Pur-Eco Chemical products. These industrial cleaning products all meet the EU Ecolabel and Nordic Ecolabel requirements and are claimed to be biodegradable, and are formulated with raw materials derived from natural vegetable sources. They also sell their products in concentrate formulas, which result in less storage requirements and packaging waste for customers.

The messaging used focuses in on business customers concerns for effective products with low environmental impact that is free of toxic chemicals. Some of the key communications in their marketing materials includes:

- Highly effective "green" cleaning for a safe and healthy work environment
- Ultra Concentrated Chemicals for a "greener" future. Product concentration leads to more efficient use of chemicals and packaging material and less impact from transportation and storage.
- Reduced plastic waste by **65%**, cardboard waste by **50%**, and carbon dioxide emissions by **62%**
- Products are biodegradable and are formulated with raw materials derived from vegetable sources
- Independent certifications of reduced environmental use of third-party eco-logos

Diversey has built sustainability into the core of their product offerings. Getting third-party certifications to reinforce to their customers that they indeed have greener products is a smart practice. Their messaging covers all the most important things to their customers: safe, effective products with lower environmental impacts. The green story is convincing due to their greener product credentials, top management support, and the fact that sustainability is woven into the company's DNA.

Key Elements of Effective Green Marketing

We have seen that effective green marketing programs, whether B2C or B2B have certain key elements.

1. Greener products are woven into the business strategy.
2. Understanding customers' desires and goals and align greener products to meet these needs.
3. Clearly communicate greener characteristics with third-party certifications or company-branded programs. Use of communication tools like environmental product profiles or company-generated labels.
4. Be authentic and credible in all marketing efforts, substantiate all claims, and be transparent.
5. Sustainable branding is an enhancement to other brand qualities—the idea that it's a great product "**and**" it has these sustainable attributes. A product's greener quality should never overshadow its purpose.

We have seen in our analysis of green marketing approaches that these elements have been successfully put into practice by consumer package goods firms, food suppliers, chemical manufacturers, electronics, and many others (King 2011). The most successful green marketing programs include all of the key elements. As stated in the beginning of this chapter, a brand has to be built on the foundations of:

a. Having a credible greener product story
b. Meeting customer demands
c. Appropriately communicating the greener attributes

References

AGweb. 2016. Consumer Demand Grows for Organic Products. http://www.agweb.com/article/consumer-demand-grows-for-organic-products-naa-associated-press/ (Accessed November 27, 2016).

BASF Sustainable Solution Steering Booklet. https://www.basf.com/us/en/company/sustainability/management-and-instruments/sustainable-solution-steering.html (Accessed January 1, 2017).

Earth911. 2016. Consumers starting to demand sustainability. http://earth911.com/business-policy/consumers-starting-to-demand-sustainability-an-overview/ (Accessed November 27, 2016).

EcoWatch. 2016. 10 Big Announcements Big Business Made to Meet Consumer Demand for Green Products. http://www.ecowatch.com/10-big-announcements-big-business-made-to-meet-consumer-demand-for-gre-1882042242.html (Accessed November 27, 2016).

GE. 2009. Ecomagination (TM) 2009 Annual Report. GE, New York, pp. 2–4, 6, 10–11, 13, 27, 29, 35, 37, 40.

GE. 2016. 2015 Annual Report. GE, New York, p. 20.

GE Ecomagination Powering the Future 2014 Report. p. 12. www.gesustainability.com/2014-performance/ecomagination (Accessed January 1, 2016).

Georgetown Environmental Law Review. 2016. https://gelr.org/2016/02/01/making-the-green-by-going-green-increased-demand-for-green-products-and-the-ftcs-role-in-a-greener-future-georgetown-environmental-law-review/ (Accessed November 27, 2016).

Green Index. http://greenindex.timberland.com/product/6795R/ (Accessed December 31, 2016).

Guevarra, Leslie. 2016. Kaiser Applies New Green Scorecard to $1B Medical Supply Chain. https://www.greenbiz.com/news/2010/05/04/kaiser-applies-new-green-scorecard-medical-supply-chain (Accessed December 10, 2016).

Hollender, Jeffery, Ashley Orgain, and Ted Nunez. 2010. *Toward a Better, More Effective Brand of Green Marketing.* Sustainability Solutions Paper. Kaplan, Princeton, NJ.

Honest Tea. 2011. Our Philosophy. http://www.honesttea.com/mission/philosophy/fairtrade/ (Accessed March 6, 2011).

Honest Tea. 2016a. 2015 Mission Report. https://www.honesttea.com/about-us/our-mission/ (Accessed December 30, 2016).

Honest Tea. 2016b. Our Mission. https://www.honesttea.com/about-us/our-mission/ (Accessed December 31, 2016).

Iannuzzi, Al. 2011. Greener Products. CRC Press, Boca Raton, FL, p. 135.

Joseph, Damian. Score Two for Sustainability. *Fast Company*. 2010;54.

King, Heather. 2010. The View from the C-Suite: Diversey's Ed Lonergan. August 30. http://www.greenbiz.com/blog/2010/08/30/view-c-suite-diverseys-ed-lonergan (Accessed March 10, 2011).

Lowe's. 2010. Product Solutions. November 15. http://www.lowes.com/cd_Product+Solution_779459462_ (Accessed December 10, 2016).

Neutrogena Naturals. 2017. *Committed, Inside and Out*, available at http://www.neutrogena.com/category/cleansers/neutrogena-+naturals.do?&utm_source=google&utm_medium=cpc&utm_campaign=&utm_term=neutrogena%20naturals&utm_content=%7cmkwid%7csDz8eB3FD_dc%7cpcrid%7c168751456986&gclid=CLj99ezdodECFQREMgodXYIFaw&gclsrc=ds (Accessed January 1, 2017).

Practice Greenhealth. 2016. Kaiser Score Card. https://www.google.com/search?q=questions+from+KP+score+card&oq=questions+from+KP+score+card&aqs=chrome..69i57.6645j0j8&sourceid=chrome&ie=UTF-8 (Accessed December 10, 2016).

Promise + Progress 2016 Corporate Sustainability Report. 2016. Steelcase Inc.

Sealed Air Diversey. 2017. Sustainability at Diversey Brochure. https://sealedair.com/ (Accessed January 3, 2017).

Seventh Generation. 2016. Baby Shampoo & Wash. http://www.seventhgeneration.com/baby-shampoo-wash?v=1116 (Accessed December 11, 2016).

Sodexo. 2016. Fiscal 2016 Corporate Responsibility Report. p. 92. http://www.sodexousa.com/sites/sdxcom-us/home/corporate-responsibility/sustainable-development.html (Accessed January 3, 2017).

Steelcase Sustainability. 2017. https://www.steelcase.com/discover/steelcase/sustainability/#products_end-of-use-recycling (Accessed May 13, 2017).

Tesco. 2016. Reducing Our Impact on the Environment. https://www.tescoplc.com/tesco-and-society/sourcing-great-products/reducing-our-impact-on-the-environment/ (Accessed December 10, 2016).

The Clorox Company. 2010. 2010 Corporate Responsibility Report. Sustainability Report. The Clorox Company, Oakland, CA.

Timberland Earthkeepers® Original Leather 6-Inch Boots. 2016. https://www.timberland.com/shop/mens-earthkeepersand-174%3B-original-leather-6-inch-boots-9703b (Accessed December 31, 2016).

Timberland GT Scramble. 2016. http://greenindex.timberland.com/product/6795R/ (Accessed December 31, 2016).

Unilever. 2016. Sustainable Sourcing. https://www.unilever.com/sustainable-living/the-sustainable-living-plan/reducing-environmental-impact/sustainable-sourcing/ (Accessed December 10, 2016).

Werbach, Adam. 2009. Using Transparency to Execute You're Strategy. In Adam Werbach, Strategy for Sustainability (pp. 109–110). Harvard Business School Publishing, Boston, MA.

Williams, Freya. 2015. Green Giants. American Management Association, New York, NY, pp. 5–7, 110.

8

Marketing Green: Best Practices from OgilvyEarth

John Jowers and Ivellisse Morales

The New Marketing Brandscape

It was a different world when David Ogilvy started Ogilvy & Mather in 1948 (Figure 8.1). World War II had ended. The soldiers had returned home. Manufacturing went from churning out war equipment to producing goods that made life a little bit easier. Brands were finding new ways to meet customer's needs, and the American people responded enthusiastically. The economy was stronger than ever.

Back then, advertising was art and copy. But in the decades that followed, the world changed fast and the world of marketing changed with it. Technology transformed people's lives at a pace unlike any other time in history. Brands turned to television to reach audiences live, and in color. Families grew fast as the "baby boom" took hold. Millions moved out of the cities to buy a better life in the suburbs. European immigrants began to be outnumbered by those from Latin America and Asia. The G.I. Bill gave birth to the emergence of a middle class. The American Dream became more attainable to more people. All the while, the beliefs, attitudes, and behaviors of the American public became much less easily defined, making audience appeals an increasingly more complex task for brands and marketers.

Today's new world of marketing requires an ability to navigate fragmented media channels, infinite information, and dizzying distraction.

There are principles from the David Ogilvy era that still ring true in today's digital world. Influenced by his research work at Gallup, Ogilvy believed in the **disciplined study of the consumer**. So much emphasis was placed on this, that he was known to relentlessly pursue keen insights that would often become the focus of an entire campaign.

One of our favorite Ogilvy-isms is unequivocally simple: **Fail to understand the people you're selling to, and you'll fail to understand their needs.**

FIGURE 8.1
Photograph © Ogilvy & Mather, all rights reserved.

In today's world, a brand cannot only be useful, it has to be purposeful or it quickly moves towards becoming irrelevant.

Today's consumer demands more. Brands need to do more than simply perform a function. Brands are expected to demonstrate compassion, to be accountable, to be trustworthy, to offer a hand to help those in need, and in the best cases, change the world for the better. Brands must show the type of human qualities we expect of the people we invite into our homes. A brand cannot only be useful, it has to be purposeful or it quickly moves toward becoming irrelevant.

The pursuit of brand purpose has been the catalyst for a wave of sustainability commitments that we have seen in recent years. The impact of climate change on human health is undeniable at this point, already causing geopolitical turmoil and mass migration. Combine those real-world challenges with increased consumer demand for transparency and we see the continued elevation of sustainability as a strategic business driver.

Where we observe some of the slippage between sound sustainability strategy and weak marketplace execution tends to be with real understanding of consumers themselves. And in particular, how the context of green marketing might require us to look a little more closely at the consumer in order for us to create effective communications that will lead to positive yet sustainable business growth.

In this chapter, we will walk you through an approach to crafting a well-positioned sustainability story. As marketers, we already know part of the answer: Carefully cater to consumers. But in this journey of learning how to market green products effectively, brands must mature from simply *existing to sell* to actually *doing good by doing well*.

None of this is revolutionary, nor even particularly novel to most green marketers. There's no shortage of sustainability communications experts who will offer their secret marketing sauce or proprietary research. But our aim is not to bombard you with data points, trends, or information to justify marketing green products. While that information can be interesting, it's not always useful when faced with a new product launch or seasonal campaign. We want to apply a sustainability lens to understanding your customer that is practical, and will help you understand the motivations, beliefs, and desires that motivate people to make more sustainable.

In this chapter, we share our trusted best practices for green marketing, including:

1. *Understanding the audience* we are trying to reach, so we understand where they are coming from
2. *Learning the triggers* that drive the type of sustainable behaviors that we want to see
3. *Navigating the "Green Gap"* and finding ways to eliminate the barriers to positive green behaviors
4. *Taking a leap of courage* if you want to become a purposeful brand that stands out in the marketplace

---

At OgilvyEarth—Ogilvy's sustainability communications team—we put David Ogilvy's principles into practice to help bridge the gap between complex sustainability issues, brand purpose, and consumer needs through strategic, integrated communications.

Putting People First

In Ogilvy's white paper *Mainstream Green*, we explored the "Green Gap"—the glaring disconnect between attitude and action as it relates to sustainable behaviors, practices, and purchases in the average consumer. In our research, 80% of Americans ranked green activities, such as buying local food and recycling, of high importance, but only 50% confirmed they actually do these activities (Bennett and Williams 2011). We believe marketing can play a leading role in solving this classic conundrum, but we first need to deeply understand our customers.

Start with Humans

Who are they? Who and what are they influenced by? What do they care about? Only with a deep understanding of who you're intending to reach can you be sure you're sending the right, resonating message. Success in green marketing begins with the curiosity to tap into human behaviors, identities, experiences, values, anxieties, and joys. You have to start with humans.

We take some of our instinctive cues from human-centered design. Over the years, this problem-solving practice has risen in popularity, bringing together business strategy, innovation, and behavior science to uncover inspiring insights that lead to impactful solutions. Consumer needs can unlock answers to virtually any problem.

Empathy exercises such as immersion and in-depth interviews contextualize customers, enabling a deeper look into lifestyles, habits, beliefs, and values. These insights help identify the exact needs that will motivate and inspire consumers toward desired actions or behaviors. "Doing your homework" will help set the context for the problem waiting to be solved.

In 1943, psychologist Abraham Maslow developed a five-tier model of human needs, ranked in hierarchical order (see Figure 8.2). His theory states that human deficiencies—unmet needs and desires—motivate us subconsciously to solve and take action (Maslow 1943). The first tier—physiological needs—captures our most basic needs for survival. We need to feel safe and secure and have all of our basic needs met before we can focus on growing our interpersonal and social relationships, developing our self-esteem, and reaching self-actualization.

As green marketers, we spend a disproportionate amount of energy and dol-lars trying to change people's beliefs, values, and attitudes towards sustain-ability. **Our operating assumption is that the motivation, receptivity, and willingness to adopt green purchasing behavior depend on the customer's current needs and context**.

Inspired by Maslow's motivational theory, we developed the *Ogilvy Hierarchy of Green* as an interpretative framework that provides a human-centered lens to understanding consumer's green behaviors and needs. Determining your customer's current tier will help you position and communicate sustainability messaging more effectively.

Ogilvy's Hierarchy of Green will help activate existing green impulses hidden within your customer (Figure 8.3). We must first identify needs, which motivate behaviors, which shape beliefs and attitudes. Our research reinforces what neuroscience research and behavioral economics have shown: Attitudes and beliefs are shaped by behaviors, not the other way around.

Let's explore each tier of the hierarchy:

- *Basic Needs:* We all need food to eat, water to drink, a place to live, and other basic necessities to thrive in our day-to-day lives. It's hard to think about polar bears in the North Pole if you are trying to make

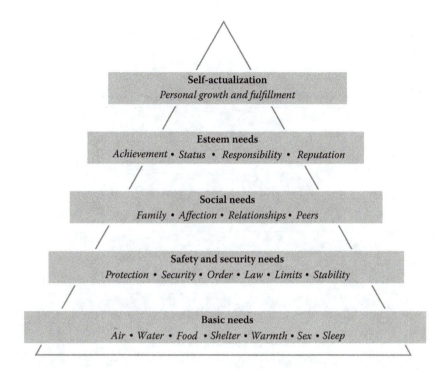

FIGURE 8.2
Maslow's Hierarchy of Needs (1943).

ends meet or have serious health issues.* Immediate basic needs also trump any aspirational motivation to move upward in the hierarchy.

- *Safety & Stability Needs:* After our basic needs are met, we seek to establish stability, which lends to a sense of security. Stable employment and retirement savings offer financial security. Health insurance, for example, provides protection and financial support. Purchasing decisions are driven by cost, convenience, and impact on personal and family health.

- *Social Needs:* Humans are conforming creatures. Whether we realize it or not, we are heavily influenced by everyone around us—family, friends, peers, and neighbors. We're also influenced by affiliation to groups, causes, and communities. Seeking acceptance, recognition,

---

* Our secondary research shows that context is critical and culturally nuanced. People of resource-constrained countries like India are observed to be naturally more green out of necessity. Poverty curbs consumption choices and behavior. Not to mention, poorer countries are disproportionately impacted by climate change (Eccleston, 2008). Wealthier countries like the United States, on the other hand, consume unsustainably. For others, sustainable behavior is law-binding. In example, the environmental stewardship of the Iroquois of North America is both a law and a collective spiritual belief in only making decisions that benefit up to seven future generations.

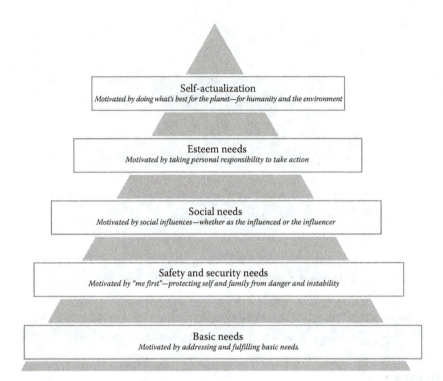

FIGURE 8.3
Ogilvy Hierarchy of Green.

and love, you're more apt to listen to recommendations and follow someone else's lead.

The above three tiers are deeply driven by personal and immediate benefits. We start to see this shift in the remaining tiers, which are motivated by future outlook and personal responsibility to preserve the planet.

- *Esteem Needs:* Once our social needs are met, we're primed to step into the tier of self-awareness. This is the tier moment where we evaluate our lifestyles, are most ready to make habit changes, and question our impact on the environment.
- *Self-Actualization:* The top tier is saved for individuals who are willing to relinquish complacency and convenience in exchange for urgent action. Lifestyle habits and changes and purchasing decisions are influenced by the existing and future climate change impacts on our planet. In our research, 71% of "Super Greens," or the top echelon of green behaviors, reported dedicating a lot of time to living a sustainable lifestyle (Bennett and Williams 2011).

Let's put the *Hierarchy of Green* into context. Imagine if Jessica, an urban city-dwelling mother of two, is in need of a new bathroom cleaner. Following our assumption, Jessica would think and behave in the following way at each tier level. You'll also find questions and considerations to guide your thinking and research.

- *Basic Needs/Are My Basic Needs Met?* "I just need a bathroom cleaner that works. All I care about is convenience and cost."
  - Questions & Considerations: How can your product address her immediate needs? Where is she most likely to go shopping for convenience and cost?
- *Safety & Stability Needs/Is This Good for Me and My Family?* "I need an affordable bathroom cleaner that is *also* healthy for my family." (In this case, green benefits become a "yes, and…")
  - Questions & Considerations: How might you make Jessica feel confident in both the price, performance, and safety of your product? How might you tap into Jessica's need to protect her family? Is your sustainable product priced within reach when compared to other products?
- *Social Needs/What Do My Family & Friends Think?* "I need a bathroom cleaner and my sister recommended this great brand that really works. My friend also recommended another more expensive, eco-friendly brand that claims to be non-toxic. I'm thinking of getting it."
  - Questions & Considerations: Who is in Jessica's ecosystem of influence and what role does she play among them? How might you leverage her influencers to tell your story? Is the performance of the product something that could be shared on social media?
- *Self-Worth Needs/Does It Make Me Feel Good?* "I need an affordable bathroom cleaner that works, is healthy for me, and is something that I can feel good about using. I am likely to research the brand's reputation, read and evaluate labels, and am more conscious of my purchasing decisions."
  - Questions & Considerations: How might your product meet Jessica's need to feel confident about her choice? How might your product tap into her growing need to be more responsible environmentally and socially? Does your product messaging make her feel good about the purchase?
- *Beyond Self-Needs/What's Best for the Planet?* "I need a bathroom cleaner that is healthy for the environment, healthy for me, and recyclable. I don't care so much about the cost, I just want the best. And if I can't find a product that fits my needs, I will make this product myself using natural ingredients."

- Questions & Considerations: How might you use information to instill trust in Jessica? How might you leverage her passion and initiative to influence others?

Grounded in a deep understanding of human psychology the *Ogilvy Hierarchy of Green* reminds us that we are all human at the end of the day. We are all driven by the same desires and needs—regardless of socioeconomic status, geographic location, or political party. We value clean air, clean water, and shelter. We value safety, love, and belonging, and respect for ourselves and others. We all value freedom, choice, and progress. These human drivers transcend time and technology. We just need a little nudge to get us there environmentally.

Nine Ways to Close the Green Gap

There's an opportunity to use behavior change as a tool to close the Green Gap. Real behavior change is rarely a one-time, one-off decision-making process. Nudging consumers toward green behaviors will shift attitudes and beliefs over time. As a brand that matters, you can cultivate a relationship that will allow you to nudge customers from one tier to the next. It's a long-term play with promising ROI.

Behavior change requires inventive ways to reach, engage, motivate, and support customers on a sustained basis. In our experience, successful behavior-change programs must be human-centered. With deep insights into your consumer's needs, you'll be able to craft communications and programing that convince individuals that:

- They have a personal stake in the issue or outcome
- Changing behavior will result in benefits they care about
- The benefits outweigh the costs
- They have the ability to change their behavior
- Services and products are available to help them

In *Mainstream Green*, we used our research to identify ways we can crack the code of one of marketing's thorniest problems: making green mainstream. Here are some of our recommended best practices to close the Green Gap (Bennett and Williams 2011).

1. Make it normal

Everybody is doing it. Normal is sustainable. Normal drives the popularity needed for a mass movement. As marketers, our predominant instincts in the sustainability space have been to market greener products as cool or different and to confer exclusive, early-adopter status on those enlightened consumers who join in. Most of those who want to go out on a green limb and self-identify as green leaders have likely already done so, but the average consumer isn't looking for things to set themselves apart from everyone else. The average consumer wants to fit in.

When it comes to driving mass behavior change, we need to make it normal. And we now know that even the *bona fide* greenies want to fit in more than we thought, so as to avoid the social stigma often associated with being an environmentally conscious consumer. Ogilvy's Rory Sutherland describes it this way, "Most people, in most fields of consumption, most of the time are Satisficers. They are simply trying to avoid making a decision that is actually bad or which might cause them to look or feel foolish. The vast bulk of money in any market at any time is in the hands of Satisficers." So how can brands market their sustainable product and service offerings in a way that makes consumers feel normal? The first principle: Make them feel like everybody's doing it.

This is what WeSpire calls "**social norming.**" WeSpire provides a Software as a Service (SAAS) platform to companies for motivating employees toward achieving sustainability impact goals. The platform's social networks, built on game mechanics, aim to infuse lightheartedness into employee engagement programs. By leveraging social networks inside a company, WeSpire found that employee engagement could not only catch on, it could drive exponential growth. When a company initially starts using their platform, the likelihood that a person will take an action is about 0.2. About a year later, a person in a workplace was five times more likely to perform that same action. "Socializers are the key to normalizing behavior, and they become your most important catalyzers in the workplace," says WeSpire CEO Susan Hunt-Stevens. "It starts out linear, but then the exponential effect takes over."

2. Make it personal

Find the "me" in green. Ask not what the consumer can do for sustainability; ask what sustainability can do for the consumer—and then show them. OgilvyEarth has long believe that we need to shift sustainability marketing from polar bears to people. Messages that are personal resonate more deeply with people than messages that are abstract, lofty, and remote. **Companies that can link their products to highly personal benefits are better positioned to succeed.**

This accounts for the success of certain sustainable product categories such as organic foods. The US organic industry posted new records in 2015, with total organic product sales hitting a new benchmark of $43.3 billion, up a robust 11% from the previous year's record level and far outstripping the overall food market's growth rate of 3%, according to the 2016 Organic Industry Survey (Organic Trade Association 2015). Consumers understand the importance of organic food because it is something that they put into their bodies and is perceived to have direct personal benefits—improved quality, taste, and "purity" owing to the absence of synthetic hormones and pesticides.

3. Make green the default

Make green more convenient and widely available. If green is the default option, people don't have to make the decision. Being green in a society where green is not widely adopted is hard, even for someone deeply rooted in the cause. Being green can mean being faced with complex choices and trade-offs in what often becomes an exhausting effort to do the right thing. But what happens if you do the hard work for people? What if you make it normal by making the better choice the default?

In the UK, Marks & Spencer first introduced a 5p charge on food carrier bags in 2008, which successfully drove a reduction of 75% in usage. Later, in 2015, after the government made the charge mandatory, retailer Tesco reported the number of plastic bags taken home by shoppers at stores in England dropped by almost 80%.

Some credit the little English town of Modbury, population 1,533, for showing the way. It had the foresight to ban plastic bags in 2007. To date, 12 US cities have a plastic bag ban or charge a plastic bag fee, including Los Angeles, San Francisco, Seattle, and New York City. In 2015, 23 states put forth 77 bills proposing a ban on plastic bags. The default movement is gaining momentum (National Conference of State Legislatures 2016).

Sometimes the best thing to do in the sustainability space is to remove the burden of complex choices from our overburdened consumers. Convenience has always sold, and making green convenient is a powerful inducement.

4. Eliminate the sustainability tax

Don't tax virtuous behavior. Governments use taxes to change behavior. Since they want fewer people to smoke, they put a hefty tax on smoking. Here's a tax break on mortgage interest paid. In Russia and other low birth-rate countries, governments encourage procreation by bestowing generous tax breaks to those willing to do their part to bring children (future taxpayers) into the world.

In the greener products market, we've got the opposite incentive going on. We tax virtuous behavior. The premium prices of many greener products on store shelves discourage purchases and perceptions. More generous government subsidies for carbon-intensive coal and oil than for clean solar and wind energy offer the same impression.

As one of the leading providers of groceries in the United States, Walmart is in a prime position to impact the price of healthy, sustainably sourced foods. The retail giant is working to reduce the price of foods made with whole grains, as well as fresh fruits and vegetables, and they are even willing to cut into their own profits to do so (Peterson 2016). While they hope that the volume of sales will make up for the reduced profits, they stand by their belief that the customer should not have to sacrifice healthier or more sustainable options based on limits of affordability.

5. Bribe shamelessly

Incentivize progress, not perfection. Gold stars, prizes, coupons—we all love rewards for our good behavior. Those new to this brave new world of greener choices may find themselves launched into a maelstrom of conflicting emotions, feeling they can never do enough and burdened with the curse of consciousness that comes with the first bite into the green apple. We can lighten this burden by offering them incremental, ongoing rewards for what they do accomplish, creating a framework that rewards individuals as they move along the green continuum. Since this is an imperfect journey we are all taking together, why not make it more enjoyable, with treats along the way?

RecycleBank, for example, rewards consumers for recycling on an ongoing basis with redeemable "points" for a range of free or discounted products. When designing their rewards program, they made sure not to confuse the desired behavior (greener energy use) with the reward. RecycleBank isn't rewarding eco-friendly behavior solely with eco-friendly rewards. It is rewarding good behavior with normal things we all want and enjoy.

This type of approach has been so successful that new entrants are rushing to attach rewards to other everyday sustainable acts such as saving energy and water. Platforms such as EarthAid, EcoBonus, and Greenopolis all partner with businesses to put money back into consumer hands for choosing more sustainable behaviors or purchasing more sustainable products.

6. Don't stop innovating

Make better stuff. We don't like going backwards. High-performing sustainable choices are key for mass adoption. Consumers are unwilling to sacrifice quality for sustainability. And rightfully so.

Unilever's Persil Small & Mighty concentrated laundry detergent saves 35 million liters of water a year in Europe—and comes with a trusted brand name. Levi's, on the back of a legacy advertising campaign, successfully brought to market a Water<Less jeans collection, that reduced Levi's' water usage by up to 96% for some styles (Levi's 2016). Companies such as Nike saw the performance challenge as an innovation opportunity. Sustainable materials for athletic shoes led to increased comfort and performance.

Thanks to a long history of premium pricing for green, the bar for sustainable products is higher. It's not enough to perform just as well; products have to perform better. In an era when opportunities to differentiate our products and brands are increasingly hard to come by, sustainability can provide fertile ground for breakthrough innovation for those marketers brave enough to turn green into gold.

7. Lose the crunch

Drop the 'G' word. Green marketing needs to be more mainstream hip than off-the-grid hippie. Not everything sustainable needs to come in brown burlap and a kale smoothie green. We need to ditch the crunch factor of green and liberate ourselves from the stereotypes. And the best way to do it may be not to mention the "G" word at all.

Julie Gilhart, former fashion director for the uber-trendy Manhattan department store, Barneys NY, and a sustainability change-agent, describes how she couldn't understand why the first fabulous, eco-friendly goods she brought into the store weren't selling as well as other items. She decided to try a different approach and removed all reference to "eco-friendly" from the labels. Sure enough, sales of the premium-priced garments picked up. She realized her discerning shopper had been turned off by the crunchy image and inferior quality the eco-friendly label cued. But in its absence, the benefits of the eco-friendly materials and production process spoke for themselves: softer, more luxurious fabrics for a premium garment.

Chevy hit the right note in its Volt positioning—a high-performing car that just happens to be sustainable. Its tagline says it all: "It's more car than electric." We call this messaging approach "P.S.: We're sustainable." Communications should embrace the fact that sustainability is a dealmaker, not a deal-breaker, for the mainstream consumer.

8. Avoid making green girly

Turn eco-friendly into male ego-friendly. Green is not a sustainable proposition for the "manly man." Carry a tote, give up your truck, compost. It's true that the everyday domestic choices we need to make in favor of sustainability do not make the average NASCAR fan's heart race. Marlboro famously cracked this code when it replaced "Mild May" in its ads with the now-iconic Marlboro Man. This strong, silent type turned smoking filtered cigarettes into a manly man's thing almost overnight. Sustainability could use its Marlboro Man moment.

So how can we make green man-friendly? A comparison between the Prius and BMW's eco-friendly car line, Efficient Dynamics, is an edifying exercise. The Prius targets early adopters with its quirky shape, advertising humor, and focus on the environment. If you're targeting early adopters looking to telegraph their green credentials, this approach is perfect. But inherent in this campaign is the message that cars are bad and must be neutered. That approach will never win over more mainstream men who want their car to tell the world how manly and successful they are. BMW taps into this desire, focusing on masculine interests such as performance, innovation, and the new frontier in luxury driving to appeal to male consumers. Other brand examples include Patagonia and Clif Bar who leverage the love of the great outdoors to inspire surfers and snowboarders to care about compromised surf and snow.

9. Make it tangible

Localize it to make it real. Sustainability is harder to follow when you can't see the trail. Find ways to help consumers see the unseeable and calculate the crazy calculus. The line from shopping cart to the Arctic is a long one. And if the carbon footprint calculation isn't easy, even for scientists, then what should we expect from consumers at point-of-sale? We need to simplify mental accounting and translate the murky benefits of sustainability into something immediate and concrete.

Automobiles are a major source of consumer greenhouse emissions. Whether you drive 50 miles a week or 500 miles, your car still expels carbon, left unseen behind you—out of sight and out of mind.

Calculating that abstract, invisible impact isn't on the agenda for most American drivers, primarily because the consequences are far-off, indistinct, and indirect. But what if the impact of driving could be felt immediately—say, on the wallets of drivers? Currently, car insurance costs the same irrespective of mileage. What if insurance were tied to how much you drive?

The Brookings Institution found that nationally driving would likely decrease by 8%; in New York state alone, by 11.5% (The Brookings Institution 2008). Moreover, only those who drove most—one-third of drivers—would be "penalized." The other two-thirds would be rewarded. In New York, Progressive has begun to experiment with a pay-as-you-drive insurance policy, while California and Massachusetts are taking the lead as part of major climate initiatives.

By closing the feedback loop, the connection between an action—driving—and its dual impacts—on the climate and your wallet—become immediate and direct. The Prius came at it another way, giving drivers real-time, on-the-dashboard visibility into the impact of their driving decisions on MPG, along with bar charts for feedback over time. This is reckoned to be a significant contributor to the Prius' success. Tying this to dollars saved could make the tool even more motivating.

These tangible signals—financial, visible, or felt in some other way—can help consumers close the feedback loop on their purchasing decisions.

Take Courage

In the absence of courage, nothing worthwhile can be accomplished.

David Ogilvy

Courage might seem like a strange point to include, but it is so often forgotten in the malaise of back-to-back meetings, stakeholder input, and approval processes. Could it be the most overlooked step to spark change, assume a leading role, and become a memorable, meaningful brand? To put it in the words of David Ogilvy: "in the absence of courage, nothing worthwhile can be accomplished."

It can be tempting to resign ourselves to the fact that we are merely marketers, communicators, social strategists, brand storytellers. Our job is simply responding to the needs of our clients. How can we really help to drive change? Are we able to change the world for the better?

David Ogilvy wrote that advertising reflects the mores of society, but doesn't influence them. History has proven him wrong. Coca-Cola's famous "Boys on a Bench" ad placed black and white young people together, even touching, and enjoying a Coke. You've seen the ad. Now look closer. It's 1969, and these boys are sitting, integrated, on a segregation bench. That image helped normalize something that was once forbidden. Our society was becoming more—not less—diverse. This was a brave decision at the time, but a smart decision based on the way the world was changing. Reversing course on inclusive advertising would have alienated the majority-in-the-making. Today, a brand could lose the trust of a whole generation.

Our work does influence the mores of society, and it is our duty to our brands and our clients to do so the right way. Climate change is real. Pandering to those who would wish it away is unsustainable, not just for the environment, but for business itself.

There will be those in our own companies and agencies who will urge what seems to be a safer course. We will hear from those who think sustainability isn't a winning message, who think that we ought to bury climate change and placate the skeptical voices in our culture.

We must do the exact opposite.

In business, our view is the long one. We work for quarterly results and plan for long-term performance. We must do the same with our marketing. It's true that we cannot fail to listen to the consumer and deeply understand them. But we must also recognize that the path toward a sustainable future is fraught with challenges. And to bring about real change requires passion, determination, but most of all, courage.

References

Bennett, Graceann and Freya Williams. Mainstream Green. *The Red Papers.* 2011,4,72–90.

Eccleston, Paul. 2008. Poorer Nations Care More for the Environment. *The Telegraph.* http://www.telegraph.co.uk/news/earth/earthnews/3341660/Poorer-nations-care-more-for-the-environment.html (Accessed January 5, 2017).

Levi's. 2016. *Water<Less.* http://store.levi.com/waterless/ (Accessed January 5, 2017).

Maslow, Abraham H. A Theory of Human Motivation. *Psychological Review* 1943;50–370–396.

National Conference of State Legislature. 2016. *State Plastic and Paper Bag Legislation.* http://www.ncsl.org/research/environment-and-natural-resources/plastic-bag-legislation.aspx (Accessed December 28, 2016).

Peterson, Hayley. 2016. *Walmart is About to Get a Lot Cheaper in One Key Area.* http://www.businessinsider.com/walmart-slashes-grocery-prices-2016-6 (Accessed January 5, 2017).

The Brookings Institution. 2008. *Pay-As-You-Drive Insurance: A Simple Way to Reduce Driving-Related Harms and Increase Equity.* https://www.brookings.edu/research/pay-as-you-drive-auto-insurance-a-simple-way-to-reduce-driving-related-harms-and-increase-equity/ (Accessed January 5, 2017).

US Organic Trade Association. 2015. *U.S. Organic Sales Post New Record of $43.3 Billion in 2015.* https://www.ota.com/news/press-releases/19031#sthash.8nicdSS6.dpuf (Accessed December 5, 2016).

9

Aspects of Green Marketing

Greenwashing

With the advent of green marketing comes a new word coined **greenwashing**. I was taught long ago that anything that has value will be mimicked with a counterfeit. This holds true with greener products. It's a lot easier to slap a label on a product that has some general term like eco-friendly or eco-conscious than to develop a truly greener product based on a strict set of criteria. Daniel Goleman, the author of *Ecological Intelligence*, states that greenwashing "pollutes the data available to consumers, gumming up marketplace efficiency by pawning off misleading information to get us to buy things that do not deliver on their promise." Additionally he states that it "undermines consumer trust, it devalues sound data, instilling doubts and cynicism in customers..."

Green-wash (green'wash', '-wôsh')—verb: the act of misleading consumers regarding the environmental practices of a company or the environmental benefits of a product or service. (TerraChoice 2011)

During training courses I have cautioned to never forget the basics. Everyone loves the accolades and notoriety given by top management for winning some external sustainability award; however, all of that is so easily forgotten when your reputation is damaged because of misleading advertising. Marketers have to be cognizant of making sure all the i's are dotted and t's are crossed when it comes to environmental claims. An increase in market share will be quickly forgotten when a major story hits the press with accusations of greenwashing.

Greenwashing can range from outrageous and bordering on the ridiculous to crossing a very fine line with good intent. Perhaps one of the funniest instances I have seen while researching this topic was an image of a Hummer, the poster child of vehicles that have really poor fuel efficiency (some models report 10 to 12 miles per gallon), painted with green logos advertising "eco-smart" and "go green" on it.

On the other hand, a tough call is the class-action lawsuits filed against SC Johnson for using "GreenList" labels on its Windex and Shout cleaning products. The lawsuit asserted that the label was misleading because it gave the impression that the product had been certified by a third party when the certification was the company's own (Vega 2010). SC Johnson was one of the pioneers of green chemistry bringing it to the mainstream and being able to demonstrate through their processes that they are making their products greener. This is one of those cases where claims are being made against a good program because they may not have had all the caveats associated with the use of GreenList as a label on packages—a good lesson for us all.

What we want to prevent is *Marketers Gone Wild.*

In a report by the marketing consulting firm OgilvyEarth titled "From Greenwash to Great," they make the case that it seems that most greenwash is the result of marketers rushing to respond to consumers' desire for greener goods and services and, in the process, falling prey to the overwhelming complexity of achieving corporate sustainability (OgilvyEarth 2011). Anyone who has worked with marketers knows that they move fast and try to get information into the consumer's hand before the competition does. By nature, marketers are extremely competitive, therefore, their exuberance has to be tempered with team efforts that work to ensure that all facts are credible and that the firm is not crossing any lines with the claims being made. Whenever I speak on preventing misleading claims, I have a saying, "What we want to prevent is **Marketers Gone Wild**." Without the proper checks, it can be very tempting to say things that make your brand look good, but are not completely steeped in fact.

One of the most prominent discussions of greenwashing has come from the consulting firm TerraChoice. A greenwashing report they published has been helpful in shining a light on the many slip-ups by companies trying to market green. As their report indicates, the more companies do green marketing the better they get at it. An evaluation of their Seven Sins of Greenwashing gives a really good idea of what not to do when communicating about sustainable products. I have found this useful to share with marketing teams so they get a good idea of what greenwash really is.

TerraChoice's Seven Sins of Greenwashing

1. *Sin of the Hidden Trade-Off:* Committed by suggesting a product is "green" based on an unreasonably narrow set of attributes without attention to other important environmental issues. Paper, for example, is not necessarily environmentally preferable just because it comes from a sustainably harvested forest. Other important environmental issues in the paper-making process, including energy, greenhouse gas emissions, and water and air pollution, may be equally or more significant.

2. *Sin of No Proof:* Committed by an environmental claim that cannot be substantiated by easily accessible supporting information or by a

reliable third-party certification. Common examples are paper products that claim various percentages of post-consumer recycled content without providing any evidence.

3. *Sin of Vagueness*: Committed by every claim that is so poorly defined or broad that its real meaning is likely to be misunderstood by the consumer. "All-natural" is an example. Arsenic, uranium, mercury, and formaldehyde are all naturally occurring, and poisonous. "All natural" isn't necessarily "green."

4. *Sin of Irrelevance*: Committed by making an environmental claim that may be truthful but is unimportant or unhelpful for consumers seeking environmentally preferable products. "CFC-free" is a common example, since it is a frequent claim despite the fact that CFCs are banned by law.

5. *Sin of Lesser of Two Evils*: Committed by claims that may be true within the product category, but that risk distracting the consumer from the greater environmental impacts of the category as a whole. Organic cigarettes might be an example of this category, as might be fuel-efficient sport-utility vehicles.

6. *Sin of Fibbing*: The least frequent sin is committed by making environmental claims that are simply false. The most common examples were products falsely claiming to be Energy Star certified or registered.

7. *Sin of Worshiping False Labels*: The Sin of Worshiping False Labels is committed by a product that, through either words or images, gives the impression of third-party endorsement where no such endorsement actually exists; fake labels, in other words (TerraChoice 2010).

The top three mistakes made by marketers are no proof, vagueness, and worshiping false labels. (TerraChoice 2010)

In evaluating the Sins of Greenwashing report, the top three mistakes made by marketers are no proof, vagueness, and worshiping false labels. Based on the high percentage of greenwashing found (over 95% of products in the study), it is a fair statement to say that it isn't easy to avoid. Though there are some marketers that are purposefully trying to mislead customers, the majority are struggling due to the newness of this type of marketing.

According to TerraChoice, one of the areas that has resulted in significant greenwash is what they call false labels. "These are labels associated with a product that are typically self-generated and intended to create the *appearance* of third-party endorsement." Most use the terms "eco," "environment," "environmentally friendly," or something similar to that. To insure that false labels are not used, it is recommended that a company use third-party standards and certifications (TerraChoice 2010).

One of the organizations that give guidance on environmental labels is the International Organization for Standardization (ISO). This is an association of standards bodies from over 160 countries that promotes the development of voluntary, consensus-based International Standards that provide solutions to global challenges. This includes environmental labeling standards. ISO standards are maintained for five different types of labels and claims ISO14020—14024. These labels align into three types of labels (ISO 2016).

Types of Eco-Labels

Type I: "A voluntary, multiple-criteria-based, third-party program that awards a license that authorizes the use of environmental labels on products indicating overall environmental preferability of a product within a particular product category based on life-cycle considerations.

Type II: Informative environmental self-declaration claims.

Type III: Voluntary programs that provide quantified environmental data of a product, under pre-set categories of parameters set by a qualified third party and based on life-cycle assessment, and verified by that or another qualified third party (Global Ecolabeling Network 2016).

According to TerraChoice, companies that have their claims endorsed by certifiers that are in line with ISO14024 are safe from greenwashing (TerraChoice 2010). This ISO standard provides guidance on developing third-party labeling programs that verify the environmental attributes of a product via a seal of approval. The standards and certifications that TerraChoice feels are the most credible are listed in Table 9.1.

I should mention that B2B marketing is not exempt from this concept. Consider the scorecards and proposal requests that are asking for all kinds of sustainability data. I am sure there will be repercussions for misleading or erroneous information given to their business customers. I would think that the same amount of care should be taken with this information as is for consumer-facing sustainability claims. At a minimum, if greenwashing is found out, damage to the business relationship and reputation of the supplier will occur.

Before leaving this topic, I would like to give a good example of how a credible and well-thought-out green marketing campaign can be developed. Hellmann's UK, a division of Unilever, wanted to make their mayonnaise more sustainable. They evaluated their raw materials and determined that eggs are one of the three main ingredients. Looking at the supply chain, it was determined that the way to improve the egg supply was to switch to free-range eggs.

This initiative resulted in many benefits: better farming practices, a quality product consumers feel good about, and a great green marketing story to tell. Hellmann's reaped some tangible benefits, positive media, and NGO attention. "Leading animal welfare organization Compassion in World Farming (CIWF) recognized this initiative—which involves sourcing 475 million free-range and barn eggs every year—by awarding Hellmann's two of its prestigious Good Egg Awards." The greener product positioning was put in advertisements as "Hellmann's Mayonnaise is now made with free range eggs" (OgilvyEarth 2011).

TABLE 9.1

TerraChoice Recommended Standards and Certifications

Terrachoice's List of Legitimate Environmental Standards and Certifications	
Biodegradable Products Institute	Natural Products Association
Chlorine Free Products Association (CFPA)	Nordic Swan
CRI Green Label	Programme for the Endorsement of Forest Certification (PEFC)
EcoCert	Rainforest Alliance
EcoLogo	Scientific Certification Systems (SCS)
ENERGY STAR	Sustainable Forestry Initiative (SFI)
OKO-TEX	Skal EKO
Fair Trade Certified	Soil Association
Forest Stewardship Council (FSC)	UL Environment Environmental Claim Validation
Green-E	UL Environment Energy Efficiency Verification
GreenGuard	USDA Organic
Green Seal	Water Sense

This is a good example of making positive changes and appropriately communicating these greener product enhancements to customers.

Regulatory Standards for Green Marketing

There are three main governmental regulatory schemes that cover environmental product-related claims, United States Federal Trade Commission (FTC) Green Guides, Canadian Competition Bureau (CSA) Environmental Claims: A Guide for Industry and Advertisers, and the UK's Department for Environment, Food and Rural Affairs (DEFRA). There have been enforcement actions taken against firms that have made misleading claims. I anticipate that enforcement actions will increase along with the many green claims being made in commerce. Any company considering making green marketing claims needs to be fully aware and compliant with these guidelines.

FTC Green Guides

The Green Guides were first issued in 1992 to give guidance to insure that environmental marketing claims are true and substantiated. The guidance they provide includes:

PART 260—GUIDES FOR THE USE OF ENVIRONMENTAL MARKETING CLAIMS

260.1 Purpose, Scope, and Structure of the Guides

260.2 Interpretation and Substantiation of Environmental Marketing Claims

The rules make deceptive acts and practices unlawful and the FTC has brought cases against companies that have crossed the line with their green marketing claims. In October 2010, proposed changes to the Guides were made to address new areas of marketing. The proposed changes would update the Guides by addressing claims about the use of "renewable materials" and "renewable energy" and give advice about carbon offset claims.

Example of FTC Enforcement Action

The FTC reached a Consent Agreement with APL, a company that markets plastic lumber, for making misleading claims about the recycled content of their product. (FTC 2016b)

Marketers would be wise to be concerned about enforcement taken by the FTC. An example of this are the charges against American Plastic Lumber (APL), a company that markets plastic lumber, regarding claims made about their products' environmental attributes that were misleading.

The FTC states that APL's ads and marketing materials "implied that its products—and the recycled plastics they contain—were made nearly all out of post-consumer recycled content, such as milk jugs and detergent bottles. In reality, the products contained less than 79% post-consumer content, on average. The FTC also charged that about 8% of APL's products contained no post-consumer recycled content at all, and nearly 7% of the products were made with only 15% post-consumer content." A consent order prohibits APL from making these environmental attribute claims which were deemed to be misleading for products and packages (FTC 2016b).

Green Certification Examples

Good Example	Bad Example

If this seal is accurate, it's **not deceptive** because it lists the specific attributes that form the basis for the product's certification.

In the FTC's Green Guides, Section 260.6, example 7, there is an example for when it is impractical to clearly list all applicable attributes adjacent to the seal itself.

This seal **may be deceptive** because it does not convey the basis for the certification. It is highly unlikely that marketers can substantiate all the attributes implied by general environmental benefit claims. That's why marketers should only use environmental certifications or seals that convey the basis for the certification.

FIGURE 9.1

Green Certification Examples. https://www.ftc.gov/news-events/press-releases/2015/09/ftc-sends-warning-letters-about-greer.

The FTC has also gone after companies that issue environmental certifications. In 2015 a letter was sent to five companies that issue certifications and to the firms that use the seals on their products, indicating that they were deceptive and may not comply with the FTC guidelines. Imagine if you were at a company using one of these seals on your product and then hearing from the FTC that the claims may be deceptive.

Below you can see a good example of what the FTC determines to be good and bad examples of a certification seal. The good example clearly labels what the improvements in the product are and the bad makes very general claims (Figure 9.1).

As we can see, the Guides are very comprehensive and not only is the FTC taking action against companies, competitors are keeping an eye out too. There have been class action law suits against firms using this law as the basis.

Canadian Competition Bureau (CSA) Environmental Claims Guide

Similar to the FTC Guides, Canada developed a Guide for making environmental claims. It gives voluntary guidance on how to comply with the Competition Act, the Consumer Packaging and Labeling Act, and the Textile Labeling Act. These are enforced by the Competition Bureau, an independent law-enforcement agency of the Government of Canada whose mission is to protect consumers from misleading advertising.

To prevent companies from greenwashing, best practices are presented for the use of labels and Type II self-declared environmental claims. These claims are usually based on a single attribute (e.g., a manufacturer's claim that a product is "biodegradable") without independent verification or certification by a third party. The guide advises that claims must be "verifiable, accurate, meaningful, and reliable if consumers are to understand the value of the environmental information they represent (e.g., their ability to protect the environment)."

Competition Act

The Competition Act is a federal law governing most business conduct in Canada. It contains both criminal and civil provisions aimed at preventing anti-competitive practices in the marketplace. The act addresses false or misleading information and deceptive marketing practices in promoting a product or service.

The Consumer Packaging and Labeling

The Consumer Packaging and Labeling Act requires that consumer products have accurate labeling to help consumers make informed purchasing decisions. "The act prohibits the making of false or misleading representations and sets out specifications for mandatory label information such as the product's name, net quantity, and dealer identity."

The Textile Labeling

The Textile Labeling Act requires that consumer textile articles bear accurate and meaningful labeling to help consumers make informed decisions. The act prohibits the false or misleading portrayals and sets mandatory label information, such as the generic name of each fiber present and the dealer's full name and postal address or identification number.

> In self-declared environmental claims, the assurance of reliability is essential. It is important that verification is properly conducted to avoid negative market effects such as trade barriers or unfair competition, which can arise from unreliable and deceptive environmental claims.
>
> (Competition Bureau Canada 2008)

An example of advice given in the guide is to avoid claims like "environmentally friendly," "ecological (eco)," and "green" due to their vagueness. For example, labeling a consumer product as "environmentally friendly" or "environmentally safe" implies that a product is environmentally benign or is environmentally beneficial" (Competition Bureau Canada 2008). Use of these very broad terms is misleading since they project an image of a product with minimal environmental impact without substantiation.

Examples of "how to" and "not to" make sustainability claims are given in the Environmental Claims Guide. Because sustainability can only be measured over a long period of time, it is difficult to verify statements. However, management systems are sometimes acceptable provided that they can be verified. An example given by the guide makes a good point and if the concept is true with any claim a company would want to make, qualify it. Claims need to be linked to the achievement and be specific.

Example:

Preferred
"This wood comes from a forest that was certified to a sustainable forest management standard [i.e., a sustainable forest management standard published by CSA, Sustainable Forestry Initiative (SFI), Forest Stewardship Council (FSC), or the Programme for the Endorsement of Forest Certification schemes (PEFC)].

Discouraged
This wood is sustainable" (Competition Bureau 2016).

Enforcement Example

The Competition Bureau issued a fine of $130,000 to EcoSmart Spas and Dynasty Spas and ordered them to stop making misleading statements that make consumers think that their Spas were Energy Star certified when they were not. The compliance order required that advertising published in stores, and on their website, that is misleading to customers is to cease and a compliance program is to be implemented to insure adherence to the regulation (Hot Tubs Works 2011). This type of public rebuke can damage a company's reputation and sway customers away from purchasing their products.

UK's Department for Environment, Food and Rural Affairs (DEFRA)

The UK, Green Claims Guidance was developed by the Department for Environment, Food and Rural Affairs (DEFRA). The Guide gives some very good direction to firms that want to make claims about the green characteristics of their products. There are three main themes: **claims should be clear, accurate, and substantiated**. These three focus areas are the essence of good green marketing. The following steps and explanations are recommended by DEFRA and make a lot of sense for any marketer that would like to make their customers aware of the environmental benefits of their products.

Step 1: Ensure the content of the claim is relevant and reflects a genuine benefit to the environment

First consider the full environmental impact of your product (and supply chain), service or organization. Check the claim is relevant to those environmental impacts, and/or your business and consumer interests. Ensure the claim does not focus on issues of low significance or importance. When comparing products, ensure the comparison is fair and relevant.

Step 2: Present the claim clearly and accurately

Ensure the claim is presented in a way that is accurate, clear, specific, and unambiguous and is easily understood by consumers. A claim shouldn't be easily misinterpreted or omit significant information. The scope of the claim should be clear; does it address the whole product or only one part. Do not use "vague, ambiguous words (e.g., "environmentally friendly") or jargon that may be easily misinterpreted or confuse consumers." All imagery must be relevant to the claim and not likely to be misinterpreted.

DEFRA Example of a Poor Practice

A product sold widely in the UK claimed: "this product is recyclable" when most areas of the UK do not have the correct infrastructure to recycle it. The term "recyclable" should only be used when there is actually an infrastructure in place that enables recycling of the product. (DEFRA 2011)

Step 3: Ensure the claim can be substantiated

There must be data to substantiate all claims (DEFRA 2011).

Green claims are enforced by the Advertising Standards Authority (ASA), the UK's regulator of advertising. They monitor claims to ensure they are "legal, decent, honest and truthful according to the advertising standards codes" (DEFRA 2011).

Misleading Claims and Enforcement

DEFRA does not enforce the regulations for making green claims other than for the European Eco-Label which they administer in the UK. The Consumer Protection from Unfair Trading Regulations (2008) requires all information to consumers to be fair and honest. Enforcement of this regulation is done by the Office of Fair Trading. Therefore, false or misleading claims can be taken up with this Office. An example of how the threat of action by the Office of Fair Trading resulted in significant changes in marketing practices is associated with sustainably sourcing seafood.

Sustainable Seafood Example

An environmental law organization, ClientEarth, conducted a study of claims made about sustainably sourced seafood in UK supermarkets. Their study indicated that "over 80% of seafood sold in the UK is by supermarkets, 25% of global fish stocks are overfished; 88% of stocks in EU waters are overfished; and 19% of stocks in EU waters are in such a bad state that scientists advise that there should be no fishing at all."

Supermarkets were making various claims indicating that the fishing methods were protective of the environment and fish stocks, such as: sustainably sourced, dolphin safe/friendly, responsibly farmed, responsibly sourced, environmentally friendly farms, and protecting the marine environment. When comparing these to the DEFRA Green Claims Guidance, it is obvious that there are some very general terms used. In fact ClientEarth determined that "32 of the 100 products reviewed, from seven supermarkets and one brand, carried claims that we consider misleading or unverified" (ClientEarth 2011).

Major retailers including Tesco, Asda, The Co-operative, Lidl, Marks & Spencer, Sainsbury's, and Waitrose were accused of carrying misleading claims on seafood items like canned tuna, haddock, and cod. Under threat of enforcement through the Office of Fair Trading, ClientEarth requested that all supermarkets having misleading claims remove them as soon as possible.

As a result of this potential enforcement, and no doubt the public relations effect of losing customers' trust, retailers and seafood providers responded. In fact, the largest supermarket in the UK, Tesco, pledged that all its canned tuna will be caught by the pole-and-line method by the end of 2012. Tesco also signed an agreement with the Sustainable Fisheries Partnership to independently review its fisheries (ClickGreen 2011).

Cause Marketing and Brand Purpose

Cause marketing is a partnership between a for-profit corporation and a nonprofit organization. These relationships are mutually beneficial in that the firm receives a halo effect when consumers choose their brand because they will be helping a worthy cause. The nonprofit benefits since they receive income to forward their groups mission. Cause marketing can be an important element in advancing a brand's sustainability story.

A report by Nielsen indicated that "55% of people are willing to pay extra for products and services from companies committed to making positive social and environmental impacts." This is even more pronounced with millennials, where 70% want firms to "do their part in addressing issues such as health, the economy, and environmental sustainability." We have seen brands respond to this by focusing on their purpose and connecting their products to causes that support the essence of the brand.

55% of people are willing to pay extra for products and services from companies committed to making positive social and environmental impacts. (Bazzarvoice 2015)

The desire to support causes with your pocketbook is not limited to the United States. Research performed by the UK-based retailer Marks & Spencer indicates that when making a purchase decision, two-thirds of their customers care about ethical choices (Goleman 2009). I believe that smart companies will consider partnering with worthy nonprofits as part of their green marketing program; when it makes sense and connects well with their brand.

However, caution should be taken when developing a cause marketing relationship. As noted by Carol Cone, a company can be considered greenwashing if the cause is not perceived as authentic and a natural fit with the brand. Whenever I have seen effective, sustainable brand cause-marketing, there is always a direct link to the cause with the purpose of the brand. This concept is supported by research performed by Kotler and Lee—they determined that full marketing benefits of a cause campaign are realized when the cause is directly related to one or more of the company's products or services (Kotler and Lee 2004).

"Cause related-marketing, as we know it, is dead. Purpose must now be engrained into the core of a company or brand's proposition. It is no longer enough to slap a ribbon on a product. It must be authentic, longterm and participatory," said Carol Cone, proclaimed "mother of cause marketing" and managing director, Brand & Corporate Citizenship, Edelman. (Edelman 2010)

The Power of Purpose

Businesses are taking notice of the benefits of tying causes and purpose into their brands. A study of business leaders indicated that 88% believe companies should have a "social purpose" and 81% see "purpose" as a business opportunity. A study indicating that companies that lead in environmental, social, and governance (ESG) have 25% higher stock value. The book "Built to Last" indicates that firms built on purpose outperformed the market 15 to 1. Based on all of these trends, business leaders would be wise to make sure they have connected their brand to a purpose that resonates with their target market.

Swedish furniture manufacturer IKEA is a prime example of purpose built into the fabric of a company; its company founder, Ingvar Kamprad, wrote a 14-page document that lays out the purpose of the company. He states that IKEA's aim is to: "Create a better everyday life for many people by offering a wide range of well-designed, functional home-furnishing products at prices so low that as many people as possible will be able to afford them."

In the food sector, Whole Foods Market is another example of purpose ingrained into the business model of the "natural"-based supermarket. John Mackey, the founder of the company speaks about his company as one that "enriches the world by its existence and brings joy, fulfillment and a sense of meaning to all that are touched by it." He even penned a book called "Conscious Capitalism" where he speaks of having a "purpose beyond profit." All this screams brand purpose, and patrons of Whole Foods are depending on these tenants to do the legwork for them to insure what they buy was produced in a responsible manner and meets high standards of sustainable production.

Further building the case for successful companies built on a purpose-driven foundation is the electric car company Tesla. Founder Elon Musk established the purpose of Tesla as, "to help expedite the move from a mine and burn economy to a solar electric economy." Revolutionizing the automotive industry with purpose as its backbone, Tesla, at the writing of this book, had a market capitalization of >\$28 billion. The purpose of this successful brand is to transform the way the global economy works—bold, and it has momentum (Williams 2015).

Build Your Brand with a Cause

Clorox did a good job of finding the nexus of brand purpose with the cause it supports. When they introduced the Green Works® brand of natural cleaners, they did it with a partnership with the Sierra Club as part of its launch. This not only gave credibility to the brands image but also enabled purchasers to support the oldest and largest grassroots environmental organization in the United States.

Another Clorox brand, Burt's Bees® initiated the "Bring Back the Bees" campaign. The initiative is not only to "raise awareness about the decrease in number of bees and pollinators, but to help rebuild a nourishing habitat for them." This connection is an excellent fit with the brand; you can't get any closer to a cause than tying it into the brand's name (Fields 2017).

Types of Cause Marketing Campaigns

Transactional campaigns unlock a business donation upon point of purchase whether via actual product sale or subsequent post-purchase consumer activity.

Digital campaigns utilize online micro-sites or social media platforms to unlock business donations and/or encourage consumer donations or other online tasks.

Licensing legally permits the use of an aspect of a nonprofit brand to be used by a company in exchange for a licensing fee.

Message-focused campaigns can take many formats but focus on utilizing business resources to share a specific cause-focused message.

Events partner a cause and a company to raise money via runs, walks, celebrations, etc., or raise awareness via clean-ups, health screenings, etc.

(Cause Marketing Forum 2010)

During a drought in Arizona, home-improvement company Home Depot initiated a program to help address water shortage. Their stores participated in a program called Use It Wisely, a water conservation campaign initiated by the Arizona Department of Water Resources. Water conservation connected well with products that minimize customers' water use and help address a community concern. As part of the campaign, Home Depot provided information on how customers can save water, such as sweeping the driveway rather than hosing it down, and installing low-flow shower-heads. This also offered an opportunity to showcase water-saving products available in stores, such as organic mulch, which reduces watering needs by 25%. Home Depot benefited by supporting this cause through connecting water conservation with their products (Kotler and Lee 2004).

Cause marketing is an important aspect of green marketing. When there is a strong connection between a brand's purpose and a cause, a sweet spot of greater customer loyalty and making the world a better place, is hit. Firms need to be cautious when initiating a cause relationship to be sure the brand is coming from a place of authenticity and is not perceived as buying its greener attributes.

Eco-Labels

Along with the increase in green marketing, there has been an escalation in the development of eco-labels (Figure 9.2). Some of them are meaningful,

FIGURE 9.2
Most commonly used Eco-labels by leading companies.

but others are not. Most consumers want to make greener product purchases, but they want it easy—so a logo indicating that a product has environmental benefits can make a consumer more confident in its product-sustainability attributes. However, because the Ecolabel Index currently lists more than 460 seals and certifications for marketing green products worldwide, that's a lot of seals out there—and a lot of confusion along with them (Idle 2016).

Eco-label proliferation makes it difficult for consumers, business customers, and marketers alike to know if a product has been legitimately improved. As an illustration, consider that most companies use some type of wood product for packaging or paper. Some of the labels available include the Green-e certified paper, Forest Stewardship Council, the Sustainable Forestry Initiative, the American Tree Farm System, Rain Forest Alliance, Recycled Paper symbol (chasing arrows), Printed With Soy Ink, and the Tropical Forest Foundation. Which one would you choose?

To add to the label confusion, there are government-issued labels like Energy Star, WaterSense, EPA Safer Choice, USDA Organic, the EU Flower, and Canadian EcoLogo. Company greener product programs can also be confused by consumers as eco-labels, such as, Home Depot's Eco Options, Staples has EcoEasy, Office Depot's Green Depot, P&G's Future Friendly, Target has Made to Matter, and Johnson & Johnson's Earthwards®. Further, there are independent companies and organizations that issue their own eco-labels or certifications, such as Underwriters Laboratories (UL), Cradle to Cradle, Green Seal, GreenGuard, and Fair Trade, to name a few.

A study conducted by the Natural Marketing Institute indicated that **the most identifiable eco-labels in the United States** are the following (% recognized listed in parenthesis):

1. Recycled logo (93%)
2. Energy Star (93%)
3. USDA Certified Organic (75%)
4. Fair Trade Certified (44%)
5. Rainforest Alliance Certified (35%)
6. Carbon Trust (24%)
7. LEED Certified (24%)
8. Green-e (19%)
9. Marine Stewardship Council (18%)
10. Sustainable Forestry Initiative (16%)

It is interesting to see that only the top three had consumer awareness above 50% (Ottman 2011). This may make marketers pause when pursuing a logo for consumer facing products. I have seen this reluctance play out in my own company. **Marketers do hesitate to add the cost of pursuing an eco-logo to the brand if they don't see a significant benefit**. However, the endorsement received from a third party does lend more credibility to a brand and may be helpful to bolster its green appeal and be a defense against green washing. According to the TerraChoice Seven Sins of Greenwashing study, products certified by an ISO 14024-based program (e.g., EcoLogo, Green Seal, and Nordic Swan) were "more than 30% sin-free (compared to the 4.4% study-wide result). In other words, good eco-labeling helps prevent (but doesn't eliminate) greenwashing" (TerraChoice 2010).

The question marketers need to ask is: Will an eco-label make a difference to my customer? If a label is desired, then which one should I pursue? One way to help make this decision is to consider the labels most widely used by the leading companies we studied in this book. A review of Figure 9.2 indicates the most commonly used eco-labels, which I noted when analyzing leading companies' green marketing programs.

Eco-labels can be helpful to demonstrate to consumers and B2B customers that environmental improvements have been built into a product. This is especially true for third-party-certified labels. However, marketers have to determine if a label will help emphasize a product's greener benefit. As in most situations, it depends on what the ultimate goal of a product is. I know that for the company for which I work, getting an eco-label makes sense for certain brands—but not all. Several factors have to be accounted for, including the actual market perception of a brand, and resonance of a specific label to the targeted customer segment. The use of company-generated labels can also be helpful to highlight to customers that their products go through rigorous "greening" before coming to market; however, there must be rock-solid data to back all sustainable attributes.

References

Bazzarvoice. 2015. How Patagonia is Using Cause Marketing to Define Their Brand and Drive Sales. http://blog.bazaarvoice.com/2015/07/07/how-patagonia-is-using-cause-marketing-to-define-their-brand-and-drive-sales/ (Accessed October 20, 2016).

Cause Marketing Forum. 2010. QuickStudy: Get up to Speed on Cause Marketing in a Hurry! http://www.causemarketingforum.com/site/c.bkLUKcOTLkK4E/b.6441799/k.81EA/Quick_Study_CM101_for_Biz.htm (Accessed March 18, 2011).

Click Green. 2011. Consumers Misled by Environmental Claims on Seafood Products. January 11. http://www.clickgreen.org.uk/analysis/business-analysis/121770-consumers-misled-by-environmental-claims-on-seafood-products.html (Accessed March 19, 2011).

ClientEarth. 2011. Environmental Claims On NGO. ClientEarth, London.

Fields, Justin. 2017. Go Behind the Scenes of Lea Michele and Burt's Bees' Brand New Partnership. http://people.com/style/exclusive-go-behind-the-scenes-of-lea-michele-and-burts-bees-brand-new-partnership/ (Accessed May 14, 2017)

Competition Bureau. 2016. Environmental Claims: A Guide for Industry and Advertisers. October 20. http://www.competitionbureau.gc.ca/eic/site/cb-bc.nsf/eng/02701.html#s1_2 (Assessed May 14, 2017).

Competition Bureau Canada. 2008. Environmental Claims: A Guide for Industry and Advertisers. Canadian Standards Association, Government, Ottawa, ON.

Department for Environment Food and Rural Affairs (DEFRA). 2011. Defra's Quick Guide to Making a Good Environmental Claim. Government, London, UK.

Edelman. 2010. Role of Citizen Consumer to Tackel Social Issues Rises. Edelman, Corporate, New York.

Federal Trade Commission (FTC). 2016a. Part 260—Guides for the Use of Environmental Marketing Claims. https://www.google.com/search?q=ftc+green+guides&oq=FTC+green&aqs=chrome.0.0j69i57j0l4.3303j0j8&sourceid=chrome&ie=UTF-8 (Accessed October 14, 2016).

FTC. 2016b. FTC approves final order green marketing case against American Plastic Lumber. https://www.ftc.gov/news-events/press-releases/2014/07/ftc-approves-final-order-green-marketing-case-against-american (Accessed October 10, 2016).

Global Ecolabeling Network. 2016. What is eco-labelling. http://www.globalecolabelling.net/what-is-eco-labelling/ (Accessed October 9, 2016).

Goleman, Daniel. 2009. The New Math. In Ecological Intelligence (p. 74). Random House, New York.

Green Certification Examples. https://www.ftc.gov/news-events/press-releases/2015/09/ftc-sends-warning-letters-about-green-certification-seals (Accessed November 6, 2016).

Hot Tubs Works. 2011. Spa Retailers Required to Stop Making False ENERGY STAR Claims. January 11. http://www.hottubworks.com/blog/spa-retailers-required-to-stop-making-false-energy-star-claims/ (Accessed March 17, 2011).

Idle, Tom. 2016. Our Use of Eco-Labels Is Set to Soar—For Products, Brands ... and People? http://www.sustainablebrands.com/news_and_views/marketing_comms/tom_idle/our_use_eco-lables_set_soar_%E2%80%93_products_brands_people (Accessed September 23, 2016).

ISO. 2016. Environmental Labels & Regulations. http://www.iso.org/iso/home/search.htm?qt=iso14024&sort=rel&type=simple&published=on (Accessed October 31, 2016).

Kotler, Philip and Nancy Lee. 2004. Best of Breed. *Stanford Social Innovation Review* Spring. http://www.ssireview.org/articles/entry/best_of_breed/. (Assessed May 14, 2017).

Ogilvyearth. 2011. From Greenwash to Great. Welcome to Ogilvyearth. http://www.ogilvyearth.com/ (Accessed March 13, 2011).

Ottman, Jacquelyn A. 2011. Establishing Credibilty and Avoiding Greenwash. In *The New Rules of Green Marketing* (p. 146). Greenleaf Publishing, Sheffield.

Selfish Giving. 2011. What is Cause Marketing. http://selfishgiving.com/cause-marketing-101/what-is-cause-marketing-2 (Accessed March 18, 2011).

TerraChoice. 2010. The Sins of Greenwashing. TerraChoice, Consulting firm, Ottawa, Canada.

TerraChoice Group Inc. 2011. The Sins of Greenwashing. http://sinsofgreenwashing. org/ (Accessed March 13, 2011).

Vega, Tanzina. 2010. Agency Seeks to Tighten Rules for "Green" Labeling. The New York Times October 6, 2010.:B4.

Williams, Freya. 2015. Green Giants. American Management Association, New York.

10

Best Practices and Conclusions

The Sustainable Brands Imperative

Interest in greener products is on the rise. We have seen, through numerous studies, that customers want products that are environmentally friendly and they want to purchase from businesses that operate in an ethical manner. According to a study conducted by Cone Communications that included nearly 10,000 citizens in nine of the largest countries in the world, 84% of global consumers say they seek out responsible products whenever possible. (SB 2016)

Why are consumers interested in purchasing greener products? Both business-to-consumer and business-to-business (B2B) consumers are responding to the global green explosion. I believe that the heightened awareness of global environmental issues like climate change, scarcity of water, air pollution, and the use of toxic chemicals is prompting a focus on doing some good with your purchase power. Companies are responding by meeting customer desire with their sustainability programs. Recent research indicates that the highest demand for greener products is coming from the countries that have the highest growth potential and the most pressing environmental problems. Countries like Brazil, China, and India have the strongest propensity for purchasing greener products. Perhaps this is due to a heightened awareness of the problems in their country and consumers' desire to purchase products that will make a difference.

I believe Walmart has been the most influential company in driving the development of more sustainable products, requiring suppliers to complete scorecards measuring their sustainability performance, and seeking greener products for their stores. Other major companies have followed suit: Kaiser Permanente, the large US-based hospital chain, developed a sustainability scorecard for their suppliers; Procter & Gamble did the same. Retailers, such as Tesco and Marks and Spencer, and home improvement and building supply chains Lowe's and Home Depot have also emphasized offering greener products to their customers and have set aggressive sustainability goals which impact companies selling in their stores.

These initiatives have influenced product development processes and are instigating competition based on sustainability. The desire for greener products touches all types of businesses and has made green marketing an imperative in B2B sales. We have seen chemical companies trying to help their customers develop more sustainable products, food industry providers focusing on fair trade and sustainably sourced raw materials, building product companies having offerings that can help customers achieve green building certifications and apparel companies greening their supply chain.

Further adding to the necessity to build sustainability into the new product development process is the fact that you are being rated. NGOs have gotten into the rating game; the Environmental Working Group has evaluated tens of thousands of products and makes it easy for the growing number of ecologically conscious consumers to choose the most wholesome product for themselves and their family. Socially responsible investment rating systems are considering the impact of products as well.

Regulatory drivers have also caused a shift in product development with the advent of environmental legislation for products. It started in the European Union with requirements on packaging design and take-back programs for electronic products. Then it moved into restrictions of certain toxic metals, flame retardants, and other chemicals. There have been many new regulatory areas with significant impact on the way business is conducted; for example, the EU REACH, RoHS, Packaging, and WEEE Directives. These product-based regulations have expanded into all regions of the world with the advent of extended producer responsibility regulations, banning of chemicals like BPA in Canada, California's Proposition 65, China REACH, Korea RoHS, Brazil packaging regulations, and the UN Globally Harmonized system for the Classification and Labeling of Chemicals (GHS), to name a few.

It seems that the regulations keep coming, and then there are the non-regulatory pressures from NGOs that are campaigning against a specific chemical or compound. We have seen effective campaigns against PVC, bisphenol-A (BPA), triclosan, phthalates, DEHP, etc. Some are based on science and some are not so scientifically based. The primary issue that companies need to be aware of is that perception is reality and that Risk = Hazard + Outrage. Therefore, monitoring emerging issues becomes a more critical part of a product stewardship program. Identifying these issues and trying to shape them with sound science and dialog is essential to the mitigation of risk. One thing is for sure, there will be more of these kinds of pressures on companies in the future.

All of these forces lead to the imperative of making and marketing greener products. So what have we learned about how leading companies respond to all of these new pressures and customer demands? We have seen that robust greener product development programs address customer, regulatory, and stakeholder demands.

Best Practices for Making Greener Products

Based on a review of leading companies, there are some common aspects to their greener product development programs. The most effective programs have **top management endorsement** and they are viewed as a business initiative rather than an environmental program. A case in point is the GE Ecomagination program. Their CEO frequently mentions this program, and it is evident that this is a company priority based on the investment in R&D ($17 billion 2005–2015).

Part of the business value of their program is connecting to customers' needs by assisting them with their sustainability initiatives. Ecomagination provides cost savings along with the environmental benefits of energy and water reduction. Samsung's CEO oversees their Green Management Committee and they also tie in cost savings of their eco-innovative programs to customers when marketing more energy-efficient products such as televisions and other electronics. Other companies have built "green" right into the foundation of their brand to the extent that it is part of the company's mission, such as Seventh Generation and Method.

Having **third-party involvement** in the development and endorsement of eco-innovation programs is also a key best practice. GE developed Ecomagination with the help of the consulting firm GreenOrder, and they have an independent advisory board that includes academics and NGOs. BASF's eco-efficiency approach and tools are verified through two independent third-party organizations, TÜVs (German technical inspection and certification organizations) and by NSF (National Sanitation Foundation). Johnson & Johnson's Earthwards® approach was developed with the assistance of the consulting firm Five Winds International and their review board includes NGOs.

The use of **scorecards, focus areas, and tools** to help identify the most important product lifecycle steps to improve upon is an important part of developing sustainable products. Examples include: the Green Index® environmental rating system used by Timberland which helps developers see a clear pathway to improving a product's environmental performance. Philips has six Green Focal Areas to make products greener: energy efficiency, packaging, toxic materials, weight, recycling & disposal, and lifetime reliability.

Samsung products are rated through the "Eco Design System (EDS)" and issue an Eco Rating. Products are put into three categories based on their eco-grading scheme; Eco-Product, Good Eco-Product, and Premium Eco-Product. Method has five key design elements: Clean—effective formulas that work, Safe—people and pet friendly, Green—safe and sustainable materials that are manufactured responsibly, Design—attractive product designs, Fragrance—use of flowers, fruits, and herbs for product scents.

Leading companies have **enterprise-wide product stewardship goals** in addition to systems and processes to green up individual brands. These goals are important to help the entire organization rally behind and drive more sustainable product design. Clorox has Eco Goals to generate one-third of growth from environmental sustainability initiatives and make sustainability improvements to 25% of their product portfolio. Philips has EcoVision which requires improving the energy efficiency of Philips products by 50%, doubling global product collection, and recycling and incorporation of recycled materials in products. Seventh Generation has set a goal to have all their paper packaging to contain 100% PCR. Unilever's has several objectives: to cut in half their environmental footprint of the making and use of their products; to help more than a billion people take action to improve their health and well-being; and to source 100% of agricultural raw materials sustainably.

Revenue from greener products is also part of some enterprise-wide goals; P&G has a $50 billion cumulative sales target for sustainable innovation products. Johnson & Johnson sets a target that 20% of revenue by 2020 will be from Earthwards® recognized products.

Use of **life-cycle analysis** or life-cycle thinking has been adopted to focus improvements on the most impactful areas. Philips has identified the key life-cycle aspects for each of their product categories to focus on making the most significant improvements. For Healthcare products, it is reducing energy consumption, weight, and dose; for Consumer Lifestyle, it is energy efficiency and closing material loops (e.g., increasing materials recycling); and for Lighting, it is energy efficiency. Method considers life cycles through their Cradle to Cradle approach. P&G looks at a product's full life cycle (raw materials, manufacturing, and product use) to help identify the most important areas to focus on. Unilever used life-cycle assessment to identify that hot-water consumption during the use of soaps and shampoos is the biggest impact area; this lead to their Turn off the Tap campaign. This program educates customers to turn off the shower while they lather to make the greatest environmental impact.

Transparency of the materials used in products has been an activity that several companies have committed to. This helps gain public trust and reinforce the credibility of the company. People trust companies that voluntarily provide more information to them than is required. Clorox is listing all ingredients on product labels for the Green Works® line of naturally derived cleaners though it is not required by law. Similarly, SC Johnson has made a commitment to make their ingredients available. Both Method and Seventh Generation have transparency commitments.

Meeting customers' needs is another key characteristic of a leader in product stewardship. Providing **end-of-life solutions** for your products is an important way to address customer requirements. This is especially true with electronic products. Apple and Samsung have established product take-back programs to facilitate recycling.

Best Practices for Making Greener Products
• Top management support and greener products are part of the business strategy
• Third-party input in developing design criteria
• Use of scorecards, focal areas, and tools to make it easier for product developers
• Enterprise-wide goals to augment individual brand improvements
• Use of life-cycle analysis or life-cycle thinking to focus on the most important impact areas
• Transparency of ingredients used in products to build more trust
• Meet customer requirements by providing end-of-life solutions for products

Best Practices for Green Marketing

The first step to effective green marketing is to have a truly greener product. In some cases, companies fear making green claims; they think it will backfire because they are not perfect. When addressing this issue, the point I try to emphasis is that **there is no such thing as a green product**. The only true green product is the one you didn't use. Every product has an environmental impact. It takes raw materials, energy to manufacture and transport the product, and there is disposal or recycling impacts. What we need to be focusing on is **greener, more sustainable products**, continuously improving, continuously reducing environmental impacts. Customers understand that you are not perfect as a company, but they do want to see that you are trying and moving in the right direction.

Customers understand that you are not perfect as a company, but they do want to see that you are trying and moving in the right direction.

Once you have a greener product, there are some common marketing approaches being taken by leading companies. One of the key components to successful green marketing is having an **effective communication method** to make it clear how your product is meeting the customers' needs. Timberland communicates with something similar to a nutritional label that indicates how the product fares in the areas of climate impact (greenhouse gas emissions), chemicals used (presence of hazardous substances), and resource consumption (use of recycled, organic, or renewable materials). Ecomagination is in itself the vehicle of communication for GE. If a product has the Ecomagination designation, it is accompanied by the actual reasons why, for example, less energy or water use.

Earthkeepers® is a branded way for Timberland to communicate the greener products that they offer. Visit their website and you can search for Earthkeepers® products and see why the shoe is more sustainable—a consumer can see the percentage of recycled content, organic materials, even information about the tannery process. Several other companies have also used their greener product development focus areas to communicate to customers. Examples include the SC Johnson GreenList® and Johnson & Johnson's Earthwards®.

Another effective method of communication is product environmental profiles. Apple maintains Product Environmental Reports for each device. The report informs customers of the impact in several areas that have been determined to be important to them: climate change, energy efficiency, material efficiency, packaging, restricted substances, and recycling. This provides the producer with a concise way to communicate product improvements as well as demonstrate their commitment to environmental protection.

Eco-labels are regularly used by leading companies to add validity to their green claims and make it easy for customers to see that the product is greener. However, this is not that easy to do since there are over 450 eco-labels and only a few are actually recognized by consumers. The eco-labels that were most used by the leading companies studied included: Recycling Symbol, Cradle to Cradle, Energy Star, Forest Stewardship Council (FSC), Sustainable Forestry Initiative (SFI), WaterSense, USDA Certified Organic, and Fair Trade. So if you think that an eco-label can help your brand, choose wisely. An illustration of good use of an eco-label is Honest Tea's use of USDA Organic, Apple's use of Energy Star, and WaterSense-labeled toilets and bathroom faucets sold at Lowe's.

Individual **company-branded greener product lines** were being employed by several leading firms. Products can be improved incrementally or developed as a new greener product from the foundation up. In both cases it is helpful to bring attention to sustainability characteristics that have been enhanced. Some of these programs have taken on eco-label–like status. Illustrations of using branded programs for products greened from the ground up are Clorox Green Works® and Timberlands Earthkeepers® line of footwear.

I have seen that Earthwards®, the branded greener product program at the company I work for, is an effective way to highlight products that have undergone eco-innovation and are greened up for better performance. These programs seem to resonate well with B2B customers because they highlight improvements that are meaningful to them, such as energy efficiency, reduced weight, more recyclable packaging, or less hazardous substances.

Cause marketing and brand purpose are important considerations for a green marketing program. With good reason too; a proof point for this is that even when the United States was in a prolonged recession, nearly three out of four Americans (72%) said that they were more likely to give their business to a company that has fair prices and supports good causes than to a company that provides deep discounts but does not contribute to good causes (Edelman 2010). Doing good through purchasing products spans the globe—customers care about making ethical choices.

There are numerous examples of using purpose and cause marketing to enhance a brand. The Sierra Club and the Green Works line of products are one such example. When Clorox wanted to bring greener products mainstream, they needed added endorsement to their natural-product claims and the Sierra Club did just that when this line of products was first introduced. Care must be taken when entering a cause relationship; it must be authentic, and not perceived as "buying" the greener-product credentials. There must be a nexus between the product and the cause it supports. Häagen-Dazs is a good example in that their ice cream relies on bees for their ingredients. Since there has been a significant reduction in the bee population, their cause is to support research to save the honey bees.

The final best practice for green marketing is a defensive maneuver: **preventing greenwashing**. Being accused of greenwashing can devastate a brand and bring a hit its reputation to a point that can be very difficult to overcome. So it makes a lot of sense to put processes in place to avoid greenwash and insure that all claims are authentic and do not overstate any environmental benefits. Some companies have relied on third-party eco-labels to bring rigor to green claims and defuse greenwashing, such as Cradle to Cradle, Energy Star, and WaterSense. Cause-based relationships with well-known NGOs also help to mitigate false or misleading claims since they must be comfortable lending their name to a brand.

The other aspect of preventing greenwashing is to insure that the sustainability claims do not overshadow the essence of the brand. Sustainability claims should complement the key aspects of the brand. One of the most significant greener products—Coldwater Tide laundry detergent—leads with the efficacy of the product and *then* mentions energy savings as a secondary message.

- 6X cleaning power.
- Tide Plus Coldwater offers 50% more energy savings when switching loads from warm to cold.

Notice that the greener benefits complement the most important product feature, cleaning power (Tide Coldwater 2016).

Green Marketing Best Practices

- Effective communication of greener characteristics to meet customers' needs through reporting environmental benefits using product stewardship focus areas and product profiles
- Use of well-known, respected eco-labels or third party's to endorse products

- Company-branded greener product lines and internal eco-label like designations
- Tapping into the brands purpose and use of appropriate cause-marketing relationships that have a direct nexus with the brand
- Preventing greenwashing by being authentic and not overstating greener product attributes

Conclusions

Things will never be the same regarding greener products; the world truly does need more sustainable brands. The stresses on the earth's systems are very evident and are growing with the expanding global middle class; people are becoming more aware of this fact. There is an increased concern about the amount of pollutants that are found in our food and even in our bloodstreams. The global demand for natural or organic products is growing and doesn't seem to be stopping.

Natural products like Method and Seventh Generation that once were considered fringe, and sold in health food stores only, are now sought after in mainstream supermarkets. The old stereotype of green products having sub-par performance has been shattered, particularly with game-changers like Green Works®, proving that large multi-nationals can develop and win in the marketplace with a naturals-based product platform.

The pull for sustainably minded products is not limited to consumer marketing. We have seen business-to-business marketing of greener products picking up pace. When chemical companies are making it a point to differentiate their products based on eco-innovation, it's time to realize that there is something here to pay attention to. Companies are seeing, through life-cycle assessments, that in many cases their biggest environmental impacts come from the acquisition of raw materials, product use, or its end of life. This new knowledge is causing a shift toward greening up products in non-traditional ways. Over recent years, sustainable innovations included cold water laundry detergents, bio-based plastics, take-back programs, and sustainably sourced raw materials, to name a few.

When you hit this *sweet spot*, of having a truly greener product that is communicated in an appropriate way, *everyone wins*.

Making products *greener* is becoming mainstream and moving toward being a requirement alongside efficacy and quality. Reduced environmental impact is being viewed as an "**and**." There are some exceptions, but greener products will not command a higher price; customers want to have a product that works **and** is greener too, but not pay more for it. This customer requirement has spanned all types of businesses; numerous examples have been discussed in this book covering a host of product categories—apparel, chemicals, building products, paper products, food, medical equipment, and packaging.

Building an eco-innovative product requires a team effort and signals from the field must be gathered to gauge customers' needs. R&D, procurement, operations, and product stewardship groups need to collaborate to build in the desired attributes. A balanced, clear communications program must be melded together by marketing and delivered by sales groups.

Care must be taken when communicating about greener products. First there must be a real authentic science-based story to be told and the message must be simple and transparent to demonstrate how the product or service helps customers with their sustainability needs.

Both customers and companies can make a difference by what they purchase and sell. Fewer resources can be used, good causes are supported, and costs are reduced. When you hit this **sweet spot**, of having a truly greener product that is communicated in an appropriate way, **everyone wins**. Customer's needs are met and brand loyalty is built.

The focus on greener products and sustainable brands is here to stay. We can expect that in the coming years that there will be a steady demand for eco-innovative products. The companies that provide these products without increasing costs will be the big winners.

References

Edelman. 2010. *Role of Citizen to Tackle Social Issues Rises*. Edelman, New York.

Sustainable Brands (SB). 2016. Study: 81% of Consumers Say They Will Make Personal Sacrifices to Address Social, Environmental Issues. http://www.sustainable-brands.com/news_and_views/stakeholder_trends_insights/sustainable_brands/study_81_consumers_say_they_will_make_ (Accessed November 11, 2016).

Tide Coldwater. 2016. Tide cold water clean liquid. http://tide.com/en-us/shop/type/liquid/tide-coldwater-clean-liquid (Accessed November 12, 2016).

Index